# SpringerBriefs in Statistics

## JSS Research Series in Statistics

The current research of statistics in Japan has expanded in several directions in line with recent trends in academic activities in the area of statistics and statistical sciences over the globe. The core of these research activities in statistics in Japan has been the Japan Statistical Society (JSS). This society, the oldest and largest academic organization for statistics in Japan, was founded in 1931 by a handful of pioneer statisticians and economists and now has a history of about 80 years. Many distinguished scholars have been members, including the influential statistician Hirotugu Akaike, who was a past president of JSS, and the notable mathematician Kiyosi Itô, who was an earlier member of the Institute of Statistical Mathematics (ISM), which has been a closely related organization since the establishment of ISM. The society has two academic journals: the Journal of the Japan Statistical Society (English Series) and the Journal of the Japan Statistical Society (Japanese Series). The membership of JSS consists of researchers, teachers, and professional statisticians in many different fields including mathematics, statistics, engineering, medical sciences, government statistics, economics, business, psychology, education, and many other natural, biological, and social sciences.

The JSS Series of Statistics aims to publish recent results of current research activities in the areas of statistics and statistical sciences in Japan that otherwise would not be available in English; they are complementary to the two JSS academic journals, both English and Japanese. Because the scope of a research paper in academic journals inevitably has become narrowly focused and condensed in recent years, this series is intended to fill the gap between academic research activities and the form of a single academic paper.

The series will be of great interest to a wide audience of researchers, teachers, professional statisticians, and graduate students in many countries who are interested in statistics and statistical sciences, in statistical theory, and in various areas of statistical applications.

More information about this series at http://www.springer.com/series/13497

Masayuki Hirukawa

# Asymmetric Kernel Smoothing

Theory and Applications in Economics and Finance

 Springer

Masayuki Hirukawa
Faculty of Economics
Ryukoku University
Kyoto
Japan

ISSN 2191-544X          ISSN 2191-5458   (electronic)
SpringerBriefs in Statistics
ISSN 2364-0057          ISSN 2364-0065   (electronic)
JSS Research Series in Statistics
ISBN 978-981-10-5465-5          ISBN 978-981-10-5466-2   (eBook)
https://doi.org/10.1007/978-981-10-5466-2

Library of Congress Control Number: 2018936652

Printed on acid-free paper

This Springer imprint is published by the registered company Springer Nature Singapore Pte Ltd. part of Springer Nature
The registered company address is: 152 Beach Road, #21-01/04 Gateway East, Singapore 189721, Singapore

# Preface

My first encounter with the nonstandard smoothing (or curve-fitting) technique by means of asymmetric kernels dates back to more than a decade ago. At that time, Nikolay (Gospodinov) and I were colleagues at Concordia University, Montreal, Canada, and we conducted a joint research project. This project specialized in pursuing a viable method of improving finite-sample properties of Stanton's (1997) nonparametric estimators for continuous-time scalar diffusion models. In the middle of the project, I happened to know the asymmetric, gamma kernel by Chen (2000) through Hagmann and Scaillet (2007). We then found superior finite-sample performance of Stanton's (1997) estimators combined with the gamma kernel, and our theoretical and empirical results were published as Gospodinov and Hirukawa (2012); see Chap. 4 for more details. Since then, I have put asymmetric kernel smoothing as one of my primary research topics and published several articles.

This book is a small collection of estimation and testing procedures using asymmetric kernels. I employ the kernels mainly for economic and financial data analyses, whereas their theoretical aspects are also of my interest. Therefore, this book is designed for a mixture of theoretical foundations and economic and financial applications of asymmetric kernel smoothing.

There are many excellent books and monographs on kernel smoothing (e.g., Silverman 1986). However, their focuses are on standard symmetric kernels. Inevitably, there are no books that pay attention to asymmetric kernels, to the best of my knowledge. I hope that this book serves as a compliment to existing books on standard kernel smoothing techniques.

This book is organized as follows. As the introduction of the book, Chap. 1 provides an informal definition and a history of asymmetric kernels and refers to the kernels that are investigated throughout. Chapters 2 and 3 deal with density estimation. While Chap. 2 discusses basic properties of the density estimators, Chap. 3 focuses on bias correction techniques in density estimation. Nonparametric regression estimation is explained in Chap. 4. Chapter 5 illustrates model specification tests smoothed by asymmetric kernels. Chapter 6 concludes with a few applications of asymmetric kernel smoothing to real data.

I would like to thank Naoto Kunitomo (Editor-in-Chief of JSS Research Series in Statistics) for encouraging me to write a book on this topic in person. In addition, he carefully read an earlier draft and gave me valuable comments and suggestions. Special thanks go to my coauthors including Benedikt Funke, Nikolay Gospodinov, and Mari Sakudo. All articles on asymmetric kernel smoothing that I coauthored with them are key ingredients of this book. I am also indebted to Yutaka Hirachi and Kavitha Palanisamy at Springer. They patiently waited for me to submit the draft. Lastly (but not least importantly), financial support from Japan Society of the Promotion of Science (grant number 15K03405) is gratefully acknowledged.

Kyoto, Japan                                                                      Masayuki Hirukawa
January 2018

# References

Chen, S.X. 2000. Probability density function estimation using gamma kernels. *Annals of the Institute of Statistical Mathematics* 52: 471–480.

Gospodinov, N., and M. Hirukawa. 2012. Nonparametric estimation of scalar diffusion models of interest rates using asymmetric kernels. *Journal of Empirical Finance* 19: 595–609.

Hagmann, M., and O. Scaillet. 2007. Local multiplicative bias correction for asymmetric kernel density estimators. *Journal of Econometrics* 141: 213–249.

Silverman, B.W. 1986. *Density Estimation for Statistics and Data Analysis*. London: Chapman and Hall.

Stanton, R. 1997. A nonparametric model of term structure dynamics and the market price of interest rate risk. *Journal of Finance* 52: 1973–2002.

# Contents

# Guide to Abbreviations and Notations

## Abbreviations

| | |
|---|---|
| ABC | Additive bias correction |
| ACD | Autoregressive conditional duration |
| B (kernel) | Beta (kernel) |
| cdf | Cumulative distribution function |
| CV | Cross-validation |
| G (kernel) | Gamma (kernel) |
| GCV | Generalized cross-validation |
| GG (kernel) | Generalized gamma (kernel) |
| *i.i.d.* | Independently and identically distributed |
| ISE | Integrated squared error |
| LL (estimator) | Local linear (estimator) |
| LMBC | Local multiplicative bias correction |
| LTBC | Local transformation bias correction |
| MB (kernel) | Modified beta (kernel) |
| MBC | Multiplicative bias correction |
| MG (kernel) | Modified gamma (kernel) |
| MISE | Mean integrated squared error |
| ML(E) | Maximum likelihood (estimate) |
| MSE | Mean squared error |
| NM (kernel) | Nakagami-$m$ (kernel) |
| NW (estimator) | Nadaraya-Watson (estimator) |
| pdf | Probability density function |
| RDD | Regression discontinuity design |
| QML(E) | Quasi-maximum likelihood (estimate) |
| SDE | Stochastic differential equation |
| SSST | Split-sample symmetry test |
| W (kernel) | Weibull (kernel) |

## Notations

| | |
|---|---|
| $B(p,q)$ | Beta function, i.e., $B(p,q) = \int_0^1 y^{p-1}(1-y)^{q-1}dy$ for $p,q > 0$ |
| $\xrightarrow{a.s.}$ | Converge almost surely to |
| $\xrightarrow{d}$ | Converge in distribution to |
| $\xrightarrow{p}$ | Converge in probability to |
| $f * g$ | Convolution of two functions $f$ and $g$, i.e., $(f * g)(x) = \int f(t)g(x-t)dt$ |
| $\Psi(a)$ | Digamma function, i.e., $\Psi(a) = d\{\log\Gamma(a)\}/da$ for $a > 0$ |
| $\|\mathbf{A}\|$ | Euclidean norm of matrix $\mathbf{A}$, i.e., $\|\mathbf{A}\| = \{\mathrm{tr}(\mathbf{A}^{\mathrm{T}}\mathbf{A})\}^{1/2}$ |
| $\Gamma(a)$ | Gamma function, i.e., $\Gamma(a) = \int_0^\infty t^{a-1}\exp(-t)dt$ for $a > 0$ |
| $\overset{iid}{\sim}$ | Independently and identically distributed as |
| $\mathbf{1}\{\cdot\}$ | Indicator function |
| $\lfloor \cdot \rfloor$ | Integer part |
| $\gamma(a,z)$ | Lower incomplete gamma function, i.e., $\gamma(a,z) = \int_0^z t^{a-1}\exp(-t)dt$ for $a,z > 0$ |
| $h^{(p)}(\cdot)$ | $p$th-order derivative of the function $h(\cdot)$ |
| $X \overset{d}{=} Y$ | A random variable $X$ obeys the distribution $Y$ |
| $X_n \asymp Y_n$ | Both $X_n/Y_n$ and $Y_n/X_n$ are bounded for all large $n$ |
| $X_n \sim Y_n$ | $X_n/Y_n \to 1$ as $n \to \infty$ |

# Chapter 1
# Asymmetric Kernels: An Introduction

This chapter presents an overview of the nonstandard smoothing technique by means of asymmetric kernels. After referring to a (relatively short) history of asymmetric kernels, we provide an informal definition and a list of the kernels. Obviously it is difficult and even uneconomical to investigate each of them fully within the limited space of this book. Instead, we concentrate on a few kernels throughout, explain why they are chosen, and illustrate their functional forms and shapes.

## 1.1 How Did Asymmetric Kernels Emerge?

### 1.1.1 Boundary Bias in Kernel Density Estimation

We should start our discussion in relation to the issue of boundary bias in kernel density estimation. The problem of estimating the unknown probability density function ("pdf") $f$ of a univariate random variable $X \in \mathbb{R}$ has been of long-lasting research interest. The most popular nonparametric estimator of the pdf is the standard kernel density estimator originated from Rosenblatt (1956) and Parzen (1962). Let $K$ be a kernel function, which is assumed to be a symmetric pdf at this moment. Examples of such kernel functions can be found, for instance, in Table 3.1 of Silverman (1986). Given $n$ observations $\{X_i\}_{i=1}^{n}$, the kernel density estimator of $f$ at a given design point $x$ using the kernel $K$ is defined as

$$\hat{f}_S (x) = \frac{1}{nh} \sum_{i=1}^{n} K \left( \frac{X_i - x}{h} \right), \tag{1.1}$$

where the smoothing parameter $h$ ($> 0$) is called the bandwidth, which controls the amount of smoothing, and the subscript "S" signifies a standard symmetric kernel. The consistency of $\hat{f}_S$ is well documented; see Parzen (1962) or Silverman (1986, Sect. 3.7.1) for a set of regularity conditions for consistency.

© The Author(s) 2018
M. Hirukawa, *Asymmetric Kernel Smoothing*, JSS Research Series
in Statistics, https://doi.org/10.1007/978-981-10-5466-2_1

The consistency result holds as long as the support of $f$ is unbounded. However, it is often the case that variables of interest take only nonnegative values. If the support of $f$ has a boundary, or more specifically, if it lies on the unit interval $[0, 1]$ or the positive half-line $\mathbb{R}_+$, then the consistency of $\hat{f}_S$ at the origin no longer holds. Because the kernel assigns positive weights outside the support when smoothing is carried out near the origin, the expected value of $\hat{f}_S(0)$ converges to $f(0)/2$ as $n \to \infty$.

## 1.1.2  Matching the Support of the Kernel with That of the Density

Perhaps Silverman (1986, Sect. 2.10) is one of the earliest studies on modifications of standard kernel density estimation in the presence of a boundary on the support. Since then, a number of remedies for the issue of boundary bias have been proposed. Examples include reflection, boundary kernel, transformation, and pseudo-data methods, to name a few. A nonexhaustive list of boundary correction methods can be found, for instance, in Zhang et al. (1999) and Karunamuni and Alberts (2005). It should be emphasized that all the methods discussed therein are built on standard symmetric kernels.

Yet another way of dealing with the boundary bias is to match the support of the kernel with that of the density to be estimated. When the support of $f$ is the positive half line, it is possible to restore consistency of the density estimate at (or in the vicinity of) the boundary by switching $K$ in (1.1) from a symmetric pdf with some pdf having support on $\mathbb{R}_+$. Because the latter never assigns positive weights on the negative half line, the resulting density estimate is free of boundary bias. The density estimator proposed by Bagai and Prakasa Rao (1995) is grounded on this idea. Because the estimator has an inferior bias convergence due to one-sided smoothing, (Guillamón et al. 1999) adopt jackknife bias correction methods to improve the bias convergence. The kernel density estimator by Abadir and Lawford (2004) may be also classified as a variant of the density estimator by Bagai and Prakasa Rao (1995). While the kernels studied in these articles are no longer symmetric, they are not categorized as asymmetric kernels because of lacking some key properties to be discussed in the next section.

## 1.1.3  Emergence of Asymmetric Kernels

It is only the late 1990s that research on asymmetric kernels began. The earliest focuses are on boundary correction of regression and density estimations with compact support. Brown and Chen (1999) approximate regression curves with support on the unit interval $[0, 1]$ using the Bernstein polynomials smoothed by a family of beta densities. As an extension of the Bernstein polynomial smoothing, Chen (1999, 2000a) propose to use a family of beta densities as kernels for density and Gasser and Müller's (1979) nonparametric regression estimators, respectively. Contrary to the chronological order in publications, the history of asymmetric kernels may have begun with regression estimation; indeed, Chen (2000a, Sect. 8) explains density estimation using the beta kernel, which is Chen's (1999) main focus. As in Bagai and Prakasa Rao (1995), Guillamón et al. (1999), and Abadir and Lawford (2004), Chen (1999, 2000a) also avoid the boundary bias by matching the support of the kernel with that of the density or regression curve. What distinguishes the latter from the former is that the latter allows kernels to vary their shapes across design points at which smoothing is made, whereas the former keeps kernels fixed everywhere.

Since the seminal work by Chen (1999, 2000a), researchers' attention has shifted to the kernels that can be applied for density estimation with support on the positive half line. The basic idea of beta kernel smoothing is extended to constructing the kernels with support on $\mathbb{R}_+$. Chen (2000b) employs gamma densities as kernels for density estimation. Jin and Kawczak (2003) investigate density estimation using the kernels constructed from log-normal and Birnbaum–Saunders densities, whereas Scaillet (2004) advocates applying inverse and reciprocal inverse Gaussian densities as kernels. Recent development and improvement of asymmetric kernels will be discussed in the next section.

### 1.1.4  Asymmetric Kernel Density Estimation as General Weight Function Estimation

We conclude this section by interpreting density estimators using asymmetric kernels in relation to Silverman's (1986) early insight. Observe that the standard kernel density estimator (1.1) can be rewritten as

$$\hat{f}_S(x) = \frac{1}{n} \sum_{i=1}^{n} K_{x,h}(X_i), \qquad (1.2)$$

where $K_{x,h}(\cdot) := (1/h) K\{(\cdot - x)/h\}$. Generalizing $K_{x,h}(\cdot)$ as the pdf that assigns most of the weight in the neighborhood of the design point $x$, Silverman (1986, Sect. 2.9) refers to (1.2) as *general weight function estimators*. The shape of the weight function $K_{x,h}(\cdot)$ may be asymmetric or even vary across the positions of $x$. Surprisingly, Silverman (1986, p. 28) does mention possibility of employing a gamma or log-normal density as the weight function when the underlying density has support on $\mathbb{R}_+$ and is right-skewed, before the advent of the gamma or log-normal kernel.

To express asymmetric kernel density estimators in the format of (1.2), let $K_{x,b}(\cdot)$ be a generic asymmetric kernel that has support either on [0, 1] or $\mathbb{R}_+$ and depends on the design point $x$ and the smoothing parameter $b(>0)$. In the literature of asymmetric kernels, the standard notation for the smoothing parameter has been $b$ since Chen (1999) rather than $h$, and this book follows the convention. Accordingly, the density estimator smoothed by the asymmetric kernel $K_{x,b}(\cdot)$ can be expressed as

$$\hat{f}(x) = \frac{1}{n} \sum_{i=1}^{n} K_{x,b}(X_i).$$

## 1.2  What Are Asymmetric Kernels?

### 1.2.1  Two Key Properties of Asymmetric Kernels

To the best of our knowledge, no formal definition of asymmetric kernels has been provided so far. Below we loosely "define" the asymmetric kernels with which we deal in this book. A weight function $K_{x,b}(\cdot)$ is said to be an *asymmetric kernel* if it possesses the following two basic properties:

**Property 1.1** *The weight function is a pdf with support either on the unit interval* [0, 1] *or on the positive half-line* $\mathbb{R}_+$.

**Property 1.2** *Both the location and shape parameters in the weight function are functions of the design point $x$ where smoothing is made and the smoothing parameter $b$.*

Property 1.1 makes nonparametric estimators smoothed by asymmetric kernels free of boundary bias, despite that they use observations with a boundary in their original scale. The property also implies that density estimates using asymmetric kernels become nonnegative everywhere. The kernels investigated by Bagai and Prakasa Rao (1995), Guillamón et al. (1999), and Abadir and Lawford (2004) also satisfy this property. However, each of these kernels $K_{x,h}(\cdot)$ is defined within the framework of the location-scale transformation, i.e., $K_{x,h}(\cdot) = (1/h) K\{(\cdot - x)/h\}$. It follows that $h$ alone controls the amount of smoothing, as is the case with standard symmetric kernels. Therefore, once the value of $h$ is fixed on the entire support, the amount of smoothing is also fixed regardless of whether the data points are dense or sparse in the vicinity of $x$.

In contrast, Property 1.2 suggests that both location and scale parameters in asymmetric kernels play a role in controlling the amount of smoothing. It follows that shapes of a given asymmetric kernel vary across design points where smoothing is made, or the amount of smoothing changes in an adaptive manner. This property will be confirmed graphically at the end of this chapter. In this respect, asymmetric kernel smoothing is reminiscent of (or may be even viewed as a version of) variable kernel (or bandwidth) methods (e.g., Abramson 1982; Silverman 1986, Sect. 2.6). The adaptive smoothing property is particularly advantageous for capturing the shapes of densities of positive random variables that are assumed to be right-skewed.

### 1.2.2 List of Asymmetric Kernels

A variety of asymmetric kernels have been proposed for the last two decades. There are a limited number of asymmetric kernels with support on [0, 1]. Examples include the beta kernel by Chen (1999) and the Gaussian-copula kernel by Jones and Henderson (2007). On the other hand, there is rich literature on asymmetric kernels with support on $\mathbb{R}_+$ (and the number of such kernels may be increasing while this book is written). Below the asymmetric kernels are classified in terms of underlying distributions.

- Gamma kernel (Chen 2000b; Jeon and Kim 2013; Igarashi and Kakizawa 2014; Malec and Schienle 2014).
- Generalized gamma kernel (Hirukawa and Sakudo 2015).
- Inverse gamma kernel (Mnatsakanov and Sarkisian 2012; Koul and Song 2013; Mousa et al. 2016; Igarashi and Kakizawa 2017).
- Inverse Gaussian and reciprocal inverse Gaussian kernels (Scaillet 2004).
- Generalized inverse Gaussian kernel (Igarashi and Kakizawa 2014).
- Birnbaum–Saunders kernel (Jin and Kawczak 2003).
- Generalized Birnbaum–Saunders kernel (Marchant et al. 2013; Saulo et al. 2013).
- Log-normal kernel (Jin and Kawczak 2003; Igarashi 2016).

It is worth emphasizing that the kernel generated from a given distribution may not be unique, regardless of whether it has support on [0, 1] or $\mathbb{R}_+$. Rather, it is possible to generate different kernels from the same distribution by changing functional forms of the shape and scale parameters. For example, the gamma kernels proposed by Igarashi and Kakizawa (2014, Sect. 4) and Malec and Schienle (2014, Sect. 2.3) can be obtained via alternative specifications of the shape parameter.

## 1.3 Which Asymmetric Kernels Are Investigated?

### 1.3.1 Scope of Asymmetric Kernels to Be Studied

It is difficult (and even uneconomical) to explain all the kernels in the list within the limited space of this book. Instead of dealing with all the kernels equally, we primarily focus on the beta and gamma kernels with support on [0, 1] and $\mathbb{R}_+$, respectively. In addition, whenever necessary, we refer to the generalized gamma kernels. The kernels can be regarded as close cousins of the gamma kernel. Like the beta and gamma kernels, asymptotic expansions of the generalized gamma kernels are boiled down to those of the gamma function. Moreover, appealing properties of the beta and gamma density estimators are inherited to the generalized gamma density estimators.

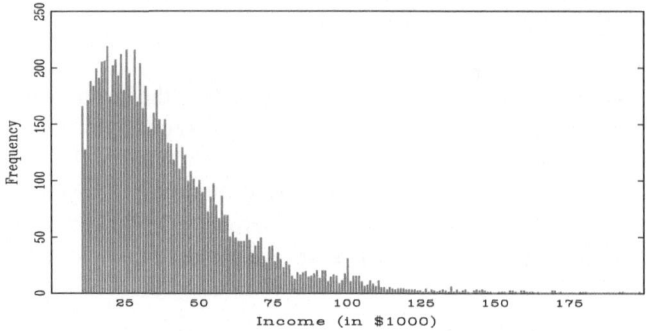

**Fig. 1.1** Histogram of US family incomes

We put a priority on these kernels because of the following reasons:

1. Empirical relevance in economics and finance;
2. Asymptotic expansion techniques; and
3. Properties of variances and boundary behavior in density estimation.

The first and second aspects will be discussed shortly. Explanations of the third aspect are deferred to Remark 2.12 and Sect. 2.5.

### 1.3.1.1  Empirical Relevance in Economics and Finance

The distributions of economic and financial variables such as wages, incomes, short-term interest rates, and insurance claims (or financial losses) can be empirically characterized by two stylized facts, namely (i) existence of a lower bound in support (most possibly at the origin) and (ii) concentration of observations near the boundary and a long tail with sparse data. Figure 1.1 is a histogram of 9275 annual family incomes in the USA. The data set is taken from Abadie (2003) and will be analyzed in detail in Chap. 6. The figure assures us of the two stylized facts. For such right-skewed distributions with a boundary, asymmetric kernels are expected to work well because of their freedom of boundary bias and adaptive smoothing property via changing shapes automatically across design points.

While asymmetric kernels are relatively new in the literature, a number of articles report favorable evidence from applying them to empirical models using the economic and financial variables. In particular, most of these empirical works apply either the beta or gamma kernel. Popularity of the kernels may be attributed to their easy implementation and attractive finite-sample properties. Table 1.1 lists nonparametric estimation problems in economics and finance using the beta or gamma kernel.

Furthermore, there are only a few works on nonparametric testing problems using asymmetric kernels, all of which will be discussed in Chap. 5. The gamma and (a few special cases of) generalized gamma kernels are exclusively considered therein.

**Table 1.1** Estimation problems in economics and finance using beta and gamma kernels

| Application | Article |
| --- | --- |
| 1. Income distribution | Bouezmarni and Scaillet (2005) [G] |
| | Hagmann and Scaillet (2007) [G] |
| 2. Actuarial loss distribution | Hagmann and Scaillet (2007) [G] |
| | Gustafsson et al. (2009) [B] |
| | Jeon and Kim (2013) [G] |
| 3. Recovery rate distribution | Renault and Scaillet (2004) [B] |
| | Hagmann et al. (2005) [B] |
| 4. Distribution of trading volumes | Malec and Schienle (2014) [G] |
| 5. Hazard rate | Bouezmarni and Rombouts (2008) [G] |
| | Bouezmarni et al. (2011) [G] |
| 6. Regression discontinuity design | Fé (2014) [G] |
| 7. Realized integrated volatility | Kristensen (2010) [B] |
| 8. Diffusion models | Gospodinov and Hirukawa (2012) [G] |

*Note* Letters "B" and "G" in brackets indicate the beta and gamma kernels, respectively

### 1.3.1.2   Asymptotic Expansion Techniques

Standard symmetric kernels are built on a set of common conditions. Typically, the kernels are symmetric, bounded, and square-integrable densities, and asymptotic properties of the estimators smoothed by the kernels can be analyzed straightforwardly by the conditions. In contrast, when delivering convergence results of asymmetric kernel estimators, we utilize mathematical tools and proof strategies that are totally different from those for symmetric kernel estimators. A tricky part is that exploring the asymptotic properties relies on kernel-specific and thus diversified approaches. To put it in another way, different asymmetric kernels require different asymptotic expansion techniques. It is uneconomical to display kernel-specific expansion techniques one-by-one. A benefit of concentrating on the beta, gamma and generalized gamma kernels is that the gamma function constitutes their core parts. Approximation techniques to the gamma function (and those to the incomplete gamma functions, which will be applied in Chap. 5) have been actively studied, and they are directly applicable to asymptotic analyses on these kernels.

## *1.3.2   Functional Forms of the Kernels*

### 1.3.2.1   Beta and Gamma Kernels

Chen (1999, 2000b) provide two definitions for each of the beta and gamma kernels. These are called "the beta (or gamma) 1 and 2 kernels", "the first and second beta (or gamma) kernels", or "the beta and modified beta (or gamma and modified gamma) kernels". This book adopts the third names throughout. Moreover, for notational conciseness, hereinafter the beta, modified beta, gamma, and modified gamma kernels are abbreviated, respectively, as the "B", "MB", "G", and "MG" kernels, whenever no confusions may occur.

The B and MB kernels are defined as the pdfs of beta distributions $Beta\{b/x + 1, (1 - x)/b + 1\}$ and $Beta\{\varrho_{b,0}(x), \varrho_{b,1}(x)\}$, respectively, for two functions $\varrho_{b,0}(x)$ and $\varrho_{b,1}(x)$ to be specified shortly. Functional forms of the kernels are given by

$$\textbf{[Beta]}: K_{B(x,b)}(u) = \frac{u^{b/x}(1-u)^{(1-x)/b}}{B\{b/x + 1, (1-x)/b + 1\}} \mathbf{1}\{u \in [0, 1]\}, \text{ and}$$

$$\textbf{[Modified Beta]}: K_{MB(x,b)}(u) = \frac{u^{\varrho_{b,0}(x)-1}(1-u)^{\varrho_{b,1}(x)-1}}{B\{\varrho_{b,0}(x), \varrho_{b,1}(x)\}} \mathbf{1}\{u \in [0, 1]\},$$

where

$$\varrho_{b,0}(x) = \begin{cases} \varrho_b(x) & \text{for } x \in [0, 2b) \\ x/b & \text{for } x \in [2b, 1] \end{cases},$$

$$\varrho_{b,1}(x) = \begin{cases} (1-x)/b & \text{for } x \in [0, 1-2b] \\ \varrho_b(1-x) & \text{for } x \in (1-2b, 1] \end{cases}, \text{ and}$$

$$\varrho_b(x) = 2b^2 + \frac{5}{2} - \sqrt{4b^4 + 6b^2 + \frac{9}{4} - x^2 - \frac{x}{b}}.$$

Likewise, the G and MG kernels are the pdfs of gamma distributions $G(x/b + 1, b)$ and $G\{\rho_b(x), b\}$ for the function $\rho_b(x)$ to be specified below. The kernels take the forms of

$$\textbf{[Gamma]}: K_{G(x,b)}(u) = \frac{u^{x/b}\exp(-u/b)}{b^{x/b+1}\Gamma(x/b + 1)} \mathbf{1}\{u \geq 0\}, \text{ and}$$

$$\textbf{[Modified Gamma]}: K_{MG(x,b)}(u) = \frac{u^{\rho_b(x)-1}\exp(-u/b)}{b^{\rho_b(x)}\Gamma\{\rho_b(x)\}} \mathbf{1}\{u \geq 0\},$$

where

$$\rho_b(x) = \begin{cases} x/b & \text{for } x \geq 2b \\ (1/4)(x/b)^2 + 1 & \text{for } x \in [0, 2b) \end{cases}.$$

Motivations to propose two modified kernels can be found in Remark 2.3. Also observe that the MB and MG kernels for interior regions are pdfs of $Beta\{b/x, (1-x)/b\}$ and $G(x/b, b)$, respectively. These densities become unbounded at the boundary. Shape parameters $\varrho_b(x)$ and $\rho_b(x)$ for boundary regions are compromises to ensure boundedness of the kernels.

### 1.3.2.2 Generalized Gamma Kernels

A family of the generalized gamma ("GG") kernels proposed by Hirukawa and Sakudo (2015) constitutes a new class of asymmetric kernels. The family consists of a specific functional form and a set of common conditions. The pdf of a GG distribution (Stacy 1962), which is also known as a special case of the Amoroso distribution (Amoroso 1925), is chosen as the functional form. The definition of the family is given below.

**Definition 1.1** (Hirukawa and Sakudo 2015, Definition 1).

Let $(\alpha, \beta, \gamma) = (\alpha_b(x), \beta_b(x), \gamma_b(x))$ be a continuous function of the design point x and the smoothing parameter b. For such $(\alpha, \beta, \gamma)$, consider the pdf of the GG distribution $GG(\alpha, \beta\Gamma(\alpha/\gamma)/\Gamma\{(\alpha+1)/\gamma\}, \gamma)$, i.e.,

$$K_{GG(x,b)}(u) = \frac{\gamma u^{\alpha-1} \exp\left[-\left\{\frac{u}{\beta\Gamma\left(\frac{\alpha}{\gamma}\right)/\Gamma\left(\frac{\alpha+1}{\gamma}\right)}\right\}^{\gamma}\right]}{\left\{\beta\Gamma\left(\frac{\alpha}{\gamma}\right)/\Gamma\left(\frac{\alpha+1}{\gamma}\right)\right\}^{\alpha}\Gamma\left(\frac{\alpha}{\gamma}\right)}\mathbf{1}\{u \geq 0\}. \tag{1.3}$$

This pdf is said to be a family of the GG kernels if it satisfies each of the following conditions:

**Condition 1.1**

$$\beta = \begin{cases} x & for\ x \geq C_1 b \\ \varphi_b(x) & for\ x \in [0, C_1 b) \end{cases}$$

for some constant $C_1 \in (0, \infty)$, where the function $\varphi_b(x)$ satisfies $C_2 b \leq \varphi_b(x) \leq C_3 b$ for some constants $0 < C_2 \leq C_3 < \infty$, and the connection between x and $\varphi_b(x)$ at $x = C_1 b$ is smooth.

**Condition 1.2** $\alpha, \gamma \geq 1$, and for $x \in [0, C_1 b)$, $\alpha$ satisfies $1 \leq \alpha \leq C_4$ for some constant $C_4 \in [1, \infty)$. Moreover, connections of $\alpha$ and $\gamma$ at $x = C_1 b$, if any, are smooth.

**Condition 1.3**

$$M_b(x) = \frac{\Gamma\left(\frac{\alpha}{\gamma}\right)\Gamma\left(\frac{\alpha+2}{\gamma}\right)}{\left\{\Gamma\left(\frac{\alpha+1}{\gamma}\right)\right\}^2} = \begin{cases} 1 + (C_5/x)b + o(b) & for\ x \geq C_1 b \\ O(1) & for\ x \in [0, C_1 b) \end{cases},$$

for some constant $|C_5| \in (0, \infty)$.

**Condition 1.4**

$$H_b(x) = \frac{\Gamma\left(\frac{\alpha}{\gamma}\right)\Gamma\left(\frac{2\alpha}{\gamma}\right)}{2^{1/\gamma}\Gamma\left(\frac{\alpha+1}{\gamma}\right)\Gamma\left(\frac{2\alpha-1}{\gamma}\right)} = \begin{cases} 1 + o(1) & if\ x/b \to \infty \\ O(1) & if\ x/b \to \kappa \in (0, \infty) \end{cases}.$$

**Condition 1.5**

$$A_{b,v}(x) = \left\{ \frac{\gamma\Gamma\left(\frac{\alpha+1}{\gamma}\right)}{\beta} \right\}^{v-1} \frac{\Gamma\left\{\frac{v(\alpha-1)+1}{\gamma}\right\}}{v^{\frac{v(\alpha-1)+1}{\gamma}}\left\{\Gamma\left(\frac{\alpha}{\gamma}\right)\right\}^{2v-1}}$$

$$= \begin{cases} V_I(v)(xb)^{(1-v)/2} + o\left(b^{(1-v)/2}\right) & if \; x/b \to \infty \\ V_B(v,\kappa)\,b^{1-v} + o\left(b^{1-v}\right) & if \; x/b \to \kappa \in (0,\infty) \end{cases}, \; v \in \mathbb{R}_+,$$

*where the subscripts "I" and "B" in the constants $V_I(v)$, $V_B(v,\kappa) \in (0,\infty)$ signify "interior" and "boundary", respectively.*

A major advantage of the family is that for each asymmetric kernel generated from this class, asymptotic properties of the kernel estimators (e.g., density and regression estimators) can be delivered by manipulating the conditions directly, as with symmetric kernels. Conditions 1.1 and 1.2 ensure that a legitimate kernel can be generated from the pdf of $GG(\alpha, \beta\Gamma(\alpha/\gamma)/\Gamma\{(\alpha+1)/\gamma\}, \gamma)$. The reason why the GG kernels are built not on the pdf of $GG(\alpha, \beta, \gamma)$ but on that of $GG(\alpha, \beta\Gamma(\alpha/\gamma)/\Gamma\{(\alpha+1)/\gamma\}, \gamma)$ will be explained in Remark 2.3. Condition 1.3 can be primarily used for the bias approximation to the GG density estimator, whereas Conditions 1.4 and 1.5 are prepared for its variance approximation. These aspects will be revisited in Remarks 2.2 and 2.5.

The family of the GG kernels embraces the following three special cases. It is easy to check that each kernel satisfies Conditions 1.1 and 1.2. The Proof of Theorem 2 in Hirukawa and Sakudo (2015) also reveals that Conditions 1.3–1.5 hold for each kernel.

**(i) MG Kernel**. Putting

$$(\alpha, \beta) = \begin{cases} (x/b, x) & for \; x \geq 2b \\ \left((1/4)(x/b)^2 + 1, x^2/(4b) + b\right) & for \; x \in [0, 2b) \end{cases}$$

and $\gamma = 1$ in (1.3) generates the MG kernel

$$K_{MG(x,b)}(u) = \frac{u^{\alpha-1}\exp\{-u/(\beta/\alpha)\}}{(\beta/\alpha)^\alpha \Gamma(\alpha)}\mathbf{1}\{u \geq 0\}.$$

It can be found that this is equivalent to the one proposed by Chen (2000b) by recognizing that $\alpha = \rho_b(x)$ and $\beta/\alpha = b$.

**(ii) Weibull Kernel**. Use the same $\beta$ as for the MG kernel but let

$$\alpha = \gamma = \begin{cases} \sqrt{2x/b} & for \; x \geq 2b \\ x/(2b) + 1 & for \; x \in [0, 2b) \end{cases}$$

in (1.3). Then, it becomes

$$K_{GG(x,b)}(u) = \frac{\alpha u^{\alpha-1} \exp\left[-\left\{\frac{u}{\beta/\Gamma(1+1/\alpha)}\right\}^{\alpha}\right]}{\{\beta/\Gamma(1+1/\alpha)\}^{\alpha}} \mathbf{1}\{u \geq 0\}.$$

The right-hand side can be rewritten as the pdf of the Weibull distribution $W(\alpha, \beta/\Gamma(1+1/\alpha))$, and thus the Weibull ("W") kernel can be defined as

$$K_{W(\alpha,\beta/\Gamma(1+1/\alpha))}(u)$$

$$= \frac{\alpha}{\beta/\Gamma(1+1/\alpha)} \left\{\frac{u}{\beta/\Gamma(1+1/\alpha)}\right\}^{\alpha-1} \exp\left[-\left\{\frac{u}{\beta/\Gamma(1+1/\alpha)}\right\}^{\alpha}\right] \mathbf{1}\{u \geq 0\}.$$

**(iii) Nakagami-$m$ Kernel.** Employ the same $(\alpha, \beta)$ as for the MG kernel, but put $\gamma = 2$ in (1.3). Then, it reduces to

$$K_{GG(x,b)}(u) = \frac{2u^{\alpha-1} \exp\left[-\{u/(\beta\Gamma(\alpha/2)/\Gamma((\alpha+1)/2))\}^2\right]}{\{\beta\Gamma(\alpha/2)/\Gamma((\alpha+1)/2)\}^{\alpha}\Gamma(\alpha/2)} \mathbf{1}\{u \geq 0\},$$

The right-hand side can be also expressed as the pdf of the Nakagami-$m$ distribution $NM\left(\alpha/2, (\alpha/2)[\beta\Gamma(\alpha/2)/\Gamma\{(\alpha+1)/2\}]^2\right)$ due to Nakagami (1943, 1960). The distribution is frequently applied in telecommunications engineering as the distribution that can describe signal intensity of shortwave fading. In the end, the Nakagami-$m$ ("NM") kernel is defined as

$$K_{NM\left(\alpha/2,(\alpha/2)[\beta\Gamma(\alpha/2)/\Gamma\{(\alpha+1)/2\}]^2\right)}(u)$$

$$= \frac{2(\alpha/2)^{\alpha/2}}{\left[(\alpha/2)\{\beta\Gamma(\alpha/2)/\Gamma((\alpha+1)/2)\}^2\right]^{\alpha/2}\Gamma(\alpha/2)} u^{2(\alpha/2)-1}$$

$$\times \exp\left[-\frac{\alpha/2}{(\alpha/2)\{\beta\Gamma(\alpha/2)/\Gamma((\alpha+1)/2)\}^2} u^2\right] \mathbf{1}\{u \geq 0\}.$$

## 1.3.3 Shapes of the Kernels

We conclude this chapter by illustrating shapes of the asymmetric kernels. Figure 1.2 plots shapes of the B (in solid lines) and MB (in dashed lines) kernels at five different design points ($x = 0.00, 0.25, 0.50, 0.75, 1.00$). Likewise, Fig. 1.3 presents shapes of the G and (three special cases of) GG kernels at four different design points ($x = 0, 1, 2, 4$). As indicated in Property 1.2, the shape of each asymmetric kernel varies according to the position at which smoothing is made; in other words, the amount of smoothing changes in a locally adaptive manner.

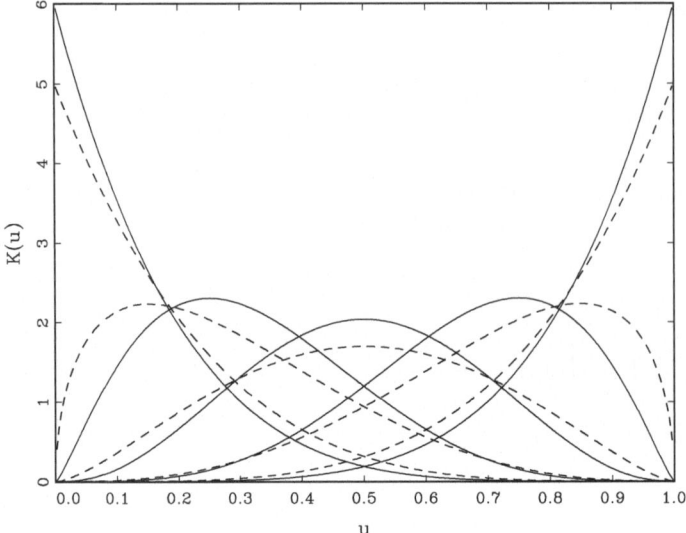

**Fig. 1.2** Shapes of the B and MB kernels for $b = 0.2$

It should be stressed that each figure is drawn with the value of the smoothing parameter fixed at $b = 0.2$. When we attempt to estimate such a density as in Fig. 1.1 using symmetric kernels, global smoothing with a fixed bandwidth may not work. If a short bandwidth is chosen so that the sharp peak near the boundary may not be missed, then the estimated right tail has spurious bumps. On the other hand, if a long bandwidth is employed to avoid the wiggly tail, then the peak is considerably smoothed away. In the end, we may be tempted to resort to variable kernel methods. In contrast, adaptive smoothing by means of asymmetric kernels can be achieved by a single smoothing parameter, which makes them much more appealing in empirical work.

A closer look at Fig. 1.2 reveals that the B and MB kernels become symmetric at $x = 0.5$, i.e., the midpoint of the unit interval. As the design point moves away from the midpoint, degrees of asymmetry increase. Also observe that when smoothing is made at each boundary, the kernels put maximum weights on the boundary. It can be found in Fig. 1.3 that when smoothing is made at the origin (Panel (a)), the NM kernel collapses to a half-normal pdf, whereas others reduce to exponential pdfs. As the design point moves away from the boundary (Panels (b)-(d)), the shape of each kernel becomes flatter and closer to symmetry.

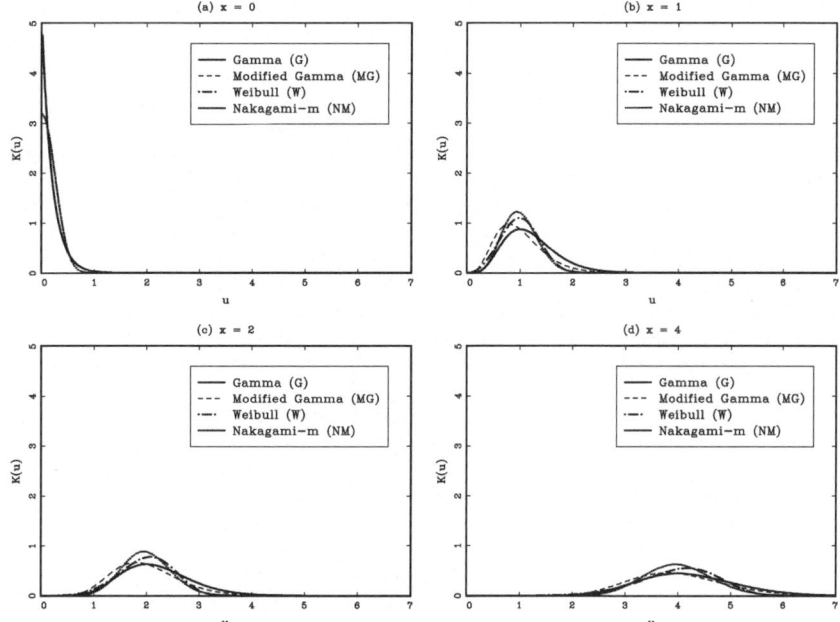

**Fig. 1.3** Shapes of the G and GG kernels for $b = 0.2$

# References

Abadie, A. 2003. Semiparametric instrumental variable estimation of treatment response models. *Journal of Econometrics* 113: 231–263.

Abadir, K.M., and S. Lawford. 2004. Optimal asymmetric kernels. *Economics Letters* 83: 61–68.

Abramson, I.S. 1982. On bandwidth variation in kernel estimates - a square root law. *Annals of Statistics* 10: 1217–1223.

Amoroso, L. 1925. Ricerche intorno alla curva dei redditi. *Annali di Matematica Pura ed Applicata, Serie IV* 10: 123–157.

Bagai, I., and B.L.S. Prakasa Rao. 1995. Kernel type density estimates for positive valued random variables. *Sankhyā: The Indian Journal of Statistics, Series A* 57: 56–67.

Bouezmarni, T., A. El Ghouch, and M. Mesfioui. 2011. Gamma kernel estimators for density and hazard rate of right-censored data. *Journal of Probability and Statistics* 2011: Article ID 937574.

Bouezmarni, T., and J.V.K. Rombouts. 2008. Density and hazard rate estimation for censored and $\alpha$-mixing data using gamma kernels. *Journal of Nonparametric Statistics* 20: 627–643.

Bouezmarni, T., and O. Scaillet. 2005. Consistency of asymmetric kernel density estimators and smoothed histograms with application to income data. *Econometric Theory* 21: 390–412.

Brown, B.M., and S.X. Chen. 1999. Beta-Bernstein smoothing for regression curves with compact support. *Scandinavian Journal of Statistics* 26: 47–59.

Chen, S.X. 1999. Beta kernel estimators for density functions. *Computational Statistics and Data Analysis* 31: 131–145.

Chen, S.X. 2000a. Beta kernel smoothers for regression curves. *Statistica Sinica* 10: 73–91.

Chen, S.X. 2000b. Probability density function estimation using gamma kernels. *Annals of the Institute of Statistical Mathematics* 52: 471–480.

Fé, E. 2014. Estimation and inference in regression discontinuity designs with asymmetric kernels. *Journal of Applied Statistics* 41: 2406–2417.

Gasser, T., and H.-G. Müller. 1979. Kernel estimation of regression functions. In *Smoothing Techniques for Curve Estimation: Proceedings of a Workshop Held in Heidelberg, April 2–4, 1979*, ed. T. Gasser and M. Rosenblatt, 23–68. Berlin: Springer-Verlag.

Gospodinov, N., and M. Hirukawa. 2012. Nonparametric estimation of scalar diffusion models of interest rates using asymmetric kernels. *Journal of Empirical Finance* 19: 595–609.

Guillamón, A., J. Navarro, and J.M. Ruiz. 1999. A note on kernel estimators for positive valued random variables. *Sankhyā: The Indian Journal of Statistics, Series A* 61: 276–281.

Gustafsson, J., M. Hagmann, J.P. Nielsen, and O. Scaillet. 2009. Local transformation kernel density estimation of loss distributions. *Journal of Business and Economic Statistics* 27: 161–175.

Hagmann, M., O. Renault, and O. Scaillet. 2005. Estimation of recovery rate densities: non-parametric and semi-parametric approaches versus industry practice. In *Recovery Risk: The Next Challenge in Credit Risk Management*, ed. E.I. Altman, A. Resti, and A. Sironi, 323–346. London: Risk Books.

Hagmann, M., and O. Scaillet. 2007. Local multiplicative bias correction for asymmetric kernel density estimators. *Journal of Econometrics* 141: 213–249.

Hirukawa, M., and M. Sakudo. 2015. Family of the generalised gamma kernels: a generator of asymmetric kernels for nonnegative data. *Journal of Nonparametric Statistics* 27: 41–63.

Igarashi, G. 2016. Weighted log-normal kernel density estimation. *Communications in Statistics - Theory and Methods* 45: 6670–6687.

Igarashi, G., and Y. Kakizawa. 2014. Re-formulation of inverse Gaussian, reciprocal inverse Gaussian, and Birnbaum-Saunders kernel estimators. *Statistics and Probability Letters* 84: 235–246.

Igarashi, G., and Y. Kakizawa. 2017. Inverse gamma kernel density estimation for nonnegative data. *Journal of the Korean Statistical Society* 46: 194–207.

Jeon, Y., and J.H.T. Kim. 2013. A gamma kernel density estimation for insurance data. *Insurance: Mathematics and Economics* 53: 569–579.

Jin, X., and J. Kawczak. 2003. Birnbaum-Saunders and lognormal kernel estimators for modelling durations in high frequency financial data. *Annals of Economics and Finance* 4: 103–124.

Jones, M.C., and D.A. Henderson. 2007. Kernel-type density estimation on the unit interval. *Biometrika* 24: 977–984.

Karunamuni, R.J., and T. Alberts. 2005. On boundary correction in kernel density estimation. *Statistical Methodology* 2: 191–212.

Koul, H.L., and W. Song. 2013. Large sample results for varying kernel regression estimates. *Journal of Nonparametric Statistics* 25: 829–853.

Kristensen, D. 2010. Nonparametric filtering of the realized spot volatility: A kernel-based approach. *Econometric Theory* 26: 60–93.

Malec, P., and M. Schienle. 2014. Nonparametric kernel density estimation near the boundary. *Computational Statistics and Data Analysis* 72: 57–76.

Marchant, C., K. Bertin, V. Leiva, and H. Saulo. 2013. Generalized Birnbaum-Saunders kernel density estimators and an analysis of financial data. *Computational Statistics and Data Analysis* 63: 1–15.

Mnatsakanov, R., and K. Sarkisian. 2012. Varying kernel density estimation on $\mathbb{R}_+$. *Statistics and Probability Letters* 82: 1337–1345.

Mousa, A.M., M.K. Hassan, and A. Fathi. 2016. A new non parametric estimator for pdf based on inverse gamma kernel. *Communications in Statistics - Theory and Methods* 45: 7002–7010.

Nakagami, M. 1943. Some statistical characters of short-wave fading (in Japanese). *Journal of the Institute of Electrical Communication Engineers of Japan* 27: 145–150.

Nakagami, M. 1960. The $m$-distribution - a general formula of intensity distribution of rapid fading. In *Statistical Methods in Radio Wave Propagation: Proceedings of a Symposium Held at the University of California, Los Angeles, June 18–20, 1958*, ed. W.C. Hoffman, 3–36. New York: Pergamon Press.

Parzen, E. 1962. On estimation of a probability density function and mode. *Annals of Mathematical Statistics* 33: 1065–1076.

Renault, O., and O. Scaillet. 2004. On the way to recovery: a nonparametric bias free estimation of recovery rate densities. *Journal of Banking and Finance* 28: 2915–2931.

Rosenblatt, M. 1956. Remarks on some nonparametric estimates of a density function. *Annals of Mathematical Statistics* 27: 832–837.

Saulo, H., V. Leiva, F.A. Ziegelmann, and C. Merchant. 2013. A nonparametric method for estimating asymmetric densities based on skewed Birnbaum-Saunders distributions applied to environmental data. *Stochastic Environmental Research and Risk Assessment* 27: 1479–1491.

Scaillet, O. 2004. Density estimation using inverse and reciprocal inverse Gaussian kernels. *Journal of Nonparametric Statistics* 16: 217–226.

Silverman, B.W. 1986. *Density Estimation for Statistics and Data Analysis*. London: Chapman and Hall.

Stacy, E.W. 1962. A generalization of the gamma distribution. *Annals of Mathematical Statistics* 33: 1187–1192.

Zhang, S., R.J. Karunamuni, and M.C. Jones. 1999. An improved estimator of the density function at the boundary. *Journal of the American Statistical Association* 94: 1231–1241.

# Chapter 2
# Univariate Density Estimation

Researchers and policy-makers are often interested in the distributions of economic and financial variables. Specifying their density functions provides natural descriptions of the distributions. This chapter investigates nonparametric density estimation using asymmetric kernels. Our main focus is on univariate density estimation. Specifically, the density estimator smoothed by the asymmetric kernel indexed by $j$ can be expressed as

$$\hat{f}_j(x) = \frac{1}{n}\sum_{i=1}^{n} K_{j(x,b)}(X_i), \tag{2.1}$$

where $j \in \{B, MB, G, MG, W, NM\}$. Whenever necessary, $j = GG$ may be used. In this case, $j$ refers to the entire family of the GG kernels. Throughout it is assumed that $\{X_i\}_{i=1}^{n}$ are *i.i.d.* random variables drawn from a univariate distribution with a pdf $f$ having support either on $[0, 1]$ or $\mathbb{R}_+$. We start from the bias and variance approximations of the density estimator (2.1) and discuss their various convergence properties. Methods of choosing the smoothing parameter are also considered. There is a list of useful formulae for asymptotic analysis in the end.

## 2.1 Bias and Variance

### 2.1.1 Regularity Conditions

Our analysis starts from bias and variance approximations. Not only do the bias and variance properties establish weak consistency of the density estimator (2.1), but also they become a building block for subsequent analyses including bias correction to be discussed in Chap. 3. Although the main focus of this section is on the bias-variance trade-off, asymptotic normality of the density estimator is straightfor-

© The Author(s) 2018

M. Hirukawa, *Asymmetric Kernel Smoothing*, JSS Research Series in Statistics, https://doi.org/10.1007/978-981-10-5466-2_2

ward to establish in a similar manner to the Proof of Theorem 5.4 in Sect. 5.5.1. Below, we present a set of regularity conditions that are standard for asymmetric kernel density estimation.

**Assumption 2.1** The second-order derivative of the pdf $f(\cdot)$ is continuous and bounded in the neighborhood of $x$.

**Assumption 2.2** The smoothing parameter $b\,(= b_n > 0)$ satisfies $b + (nb)^{-1} \to 0$ as $n \to \infty$.

### 2.1.2  Bias Approximation

The bias of the density estimator (2.1) can be approximated under Assumptions 2.1 and 2.2. It is easy to see that

$$E\left\{\hat{f}_j(x)\right\} = \int K_{j(x,b)}(u)\,f(u)\,du = E\left\{f\left(\theta_{j(x,b)}\right)\right\},\qquad(2.2)$$

where $\theta_{j(x,b)}$ is the random variable obeying the distribution with the pdf $K_{j(x,b)}(\cdot)$, i.e., $\theta_{G(x,b)} \overset{d}{=} G(x/b+1, b)$, for instance. Then, a second-order Taylor expansion of $f\left(\theta_{j(x,b)}\right)$ around $\theta_{j(x,b)} = x$ yields

$$E\left\{\hat{f}_j(x)\right\} = f(x) + E\left(\theta_{j(x,b)} - x\right) f^{(1)}(x)$$
$$+ \frac{1}{2} E\left\{\left(\theta_{j(x,b)} - x\right)^2\right\} f^{(2)}(x) + R_{\hat{f}_j(x)},$$

where

$$R_{\hat{f}_j(x)} := \int \left\{f^{(2)}(\xi) - f^{(2)}(x)\right\} (\xi - x)^2 K_{j(x,b)}(u)\,du$$

is the remainder term with $\xi = \alpha u + (1 - \alpha)x$ for some $\alpha \in (0, 1)$, which is shown to be of smaller order in magnitude. Finally, evaluating the moments $E\left(\theta_{j(x,b)} - x\right)$ and $E\left\{\left(\theta_{j(x,b)} - x\right)^2\right\}$ by means of formulae given in Sect. 2.8 yields the leading bias term. It can be found that the leading bias of $\hat{f}_j(x)$ is $O(b)$. In particular,

$$Bias\left\{\hat{f}_j(x)\right\} = \mathcal{B}_{1,j}(x, f)\,b + o(b),$$

for some kernel-specific coefficient $\mathcal{B}_{1,j}(x, f)$ that depends on the design point $x$ and derivatives of the pdf $f$. Forms of $\mathcal{B}_{1,j}(x, f)$ are given in Table 2.1 below. Notice that the table also presents forms of $\mathcal{B}_{2,j}(x, f)$, the kernel-specific coefficient on the $O(b^2)$ bias term, except the W kernel. They are required for bias approximations in a variety of bias-corrected estimators. Section 3.2 and Remark 3.5 will explain how $\mathcal{B}_{2,j}(x, f)$ appears in bias approximations of the bias-corrected density estimators and why $\mathcal{B}_{2,W}(x, f)$ is not listed in Table 2.1, respectively.

**Table 2.1** Explicit forms of $\mathcal{B}_{1,j}(x,f)$ and $\mathcal{B}_{2,j}(x,f)$

| Kernel ($j$) | | $\mathcal{B}_{1,j}(x,f)$ | $\mathcal{B}_{2,j}(x,f)$ |
|---|---|---|---|
| B | for $x \in [0,1]$ | $(1-2x)f^{(1)}(x) + \frac{x}{2}(1-x)f^{(2)}(x)$ | $-2(1-2x)f^{(1)}(x) + \frac{1}{2}(11x^2 - 11x + 2)f^{(2)}(x) +$ $\frac{5}{6}x(1-x)(1-2x)f^{(3)}(x) + \frac{x^2}{8}(1-x)^2 f^{(4)}(x)$ |
| MB | for $x \in [2b, 1-2b]$ | $\frac{x}{2}(1-x)f^{(2)}(x)$ | $-\frac{x}{2}(1-x)f^{(2)}(x) + \frac{x}{3}(1-x)(1-2x)f^{(3)}(x) + \frac{x^2}{8}(1-x)^2 f^{(4)}(x)$ |
| | for $x \in [0, 2b)$ | $\varsigma_b(x)f^{(1)}(x)$ | $\frac{1}{2}\left\{\varsigma_b^2(x) + \varsigma_b(x) + \frac{x}{b}\right\}f^{(2)}(x)$ |
| | for $x \in (1-2b, 1]$ | $-\varsigma_b(1-x)f^{(1)}(x)$ | $\frac{1}{2}\left\{\varsigma_b^2(1-x) + \varsigma_b(1-x) + \frac{1-x}{b}\right\}f^{(2)}(x)$ |
| G | for $x \ge 0$ | $f^{(1)}(x) + \frac{x}{2}f^{(2)}(x)$ | $f^{(2)}(x) + \frac{5}{6}xf^{(3)}(x) + \frac{x^2}{8}f^{(4)}(x)$ |
| MG | for $x \ge 2b$ | $\frac{x}{2}f^{(2)}(x)$ | $\frac{x}{3}f^{(3)}(x) + \frac{x^2}{8}f^{(4)}(x)$ |
| | for $x \in [0, 2b)$ | $\xi_b(x)f^{(1)}(x)$ | $\frac{1}{2}\left\{\xi_b^2(x) + \xi_b(x) + \frac{x}{b}\right\}f^{(2)}(x)$ |
| W | for $x \ge 2b$ | $\frac{\pi^2}{24}xf^{(2)}(x)$ | — |
| | for $x \in [0, 2b)$ | $\xi_b(x)f^{(1)}(x)$ | — |
| NM | for $x \ge 2b$ | $\frac{x}{4}f^{(2)}(x)$ | $\frac{1}{8}\left[\frac{1}{1}f^{(1)}(x) + \frac{x}{3}f^{(2)}(x) + \frac{x^2}{4}f^{(3)}(x)\right]$ |
| | for $x \in [0, 2b)$ | $\xi_b(x)f^{(1)}(x)$ | $\frac{1}{2}\left[\left\{\xi_b(x) + \frac{x}{b}\right\}^2 \frac{\Gamma\!\left(\frac{\varsigma_b(x)+x/b}{b}\right)\Gamma\!\left(\frac{\varsigma_b(x)+x/b}{2}+1\right)}{\Gamma^2\!\left(\frac{\varsigma_b(x)+x/b+1}{2}\right)} - 2\left(\frac{x}{b}\right)\xi_b(x) + \left(\frac{x}{b}\right)^2\right]f^{(2)}(x)$ |

*Note* $\varsigma_b(x) := (1-x)\{\varrho_b(x) - x/b\}/[1 + b\{\varrho_b(x) - x/b\}]$ and $\xi_b(x) := \{(1/2)(x/b) - 1\}^2$.

*Remark 2.1* As documented in Lemma 6 of Bouezmarni and Rolin (2003), both $\hat{f}_B(x)$ and $\hat{f}_{MB}(x)$ are unbiased for $\mathbf{1}\{0 \le x \le 1\}$, i.e., the pdf of $U[0, 1]$. This is obvious from (2.2).

*Remark 2.2* For $\hat{f}_{GG}(x)$, Condition 1.3 serves as a high-level assumption, and it leads to the leading bias coefficient

$$\mathcal{B}_{1,GG}(x, f) = \begin{cases} (C_5/2) x f^{(2)}(x) & \text{for } x \ge C_1 b \\ \xi_b(x) f^{(1)}(x) & \text{for } x \in [0, C_1 b) \end{cases}, \tag{2.3}$$

where

$$\xi_b(x) := \frac{\varphi_b(x) - x}{b} = O(1).$$

For each of the MG, W, and NM kernels, $C_1 = 2$ and $\varphi_b(x) = x^2/(4b) + b$ so that $\xi_b(x) = \{(1/2)(x/b) - 1\}^2$. On the other hand, as indicated in Table 2.1, the values of $C_5$ differ; to be more precise, $(C_{5,MG}, C_{5,W}, C_{5,NM}) = (1, \pi^2/12, 1/2)$.

*Remark 2.3* It is also worth explaining why the MB and MG are proposed as modifications of the B and G kernels, respectively. As in Table 2.1, $\mathcal{B}_{1,B}(x, f)$ and $\mathcal{B}_{1,G}(x, f)$, the leading bias coefficients of the B and G density estimators, include $f^{(1)}(x)$ as well as $f^{(2)}(x)$. This comes from the fact that $E(\theta_{B(x,b)}) = (x + b)/(1 + 2b) = x + O(b)$ and $E(\theta_{G(x,b)}) = x + b$ by (2.13) and (2.14), respectively. Because these expected values are not exactly $x$, extra terms with $f^{(1)}(x)$ are included in the corresponding bias coefficients. Modified versions of the kernels, namely the MB and MG kernels, are designed to eliminate the term involving $f^{(1)}(x)$ as long as $x$ is away from the boundary. Indeed, $\theta_{MB(x,b)} \stackrel{d}{=} Beta\{x/b, (1 - x)/b\}$ for $x \in [2b, 1 - 2b]$ and $\theta_{MG(x,b)} \stackrel{d}{=} G(x/b, b)$ for $x \ge 2b$. It follows again from (2.13) and (2.14) that $E(\theta_{MB(x,b)}) = E(\theta_{MG(x,b)}) = x$, which leads to the bias coefficient without $f^{(1)}(x)$. Similarly, we can also understand why not $GG(\alpha, \beta, \gamma)$ but $GG(\alpha, \beta\Gamma(\alpha/\gamma)/\Gamma\{(\alpha + 1)/\gamma\}, \gamma)$ is chosen as the distribution that generates the GG kernels. For $\theta_{GG(x,b)} \stackrel{d}{=} GG(\alpha, \beta\Gamma(\alpha/\gamma)/\Gamma\{(\alpha + 1)/\gamma\}, \gamma)$, $E(\theta_{GG(x,b)}) = \beta$ by (2.15). Therefore, $E(\theta_{GG(x,b)}) = x$ for $x \ge 2b$, which again yields the bias coefficient with only $f^{(2)}(x)$. The GG kernels are designed to inherit all appealing properties that the MG kernel possesses, and this result is one of such properties.

*Remark 2.4* We can compare bias approximations of $\hat{f}_j(x)$ and $\hat{f}_S(x)$, the density estimator using a nonnegative symmetric kernel defined in (1.1). It is well known that the bias expansion of $\hat{f}_S(x)$ takes a much simpler form of

$$Bias\{\hat{f}_S(x)\} = \frac{\mu_2}{2} f^{(2)}(x) h^2 + o(h^2),$$

where $\mu_2 = \int_{-\infty}^{\infty} u^2 K(u) du$. Inclusion of $f^{(1)}(x)$ in dominant bias terms of some asymmetric kernel density estimators reflects that while odd-order moments of symmetric kernels are exactly zero, first-order moments of asymmetric kernels around the design point $x$ are $O(b)$.

### 2.1.3 Variance Approximation

An approximation to the variance of $\hat{f}_j(x)$ under Assumptions 2.1 and 2.2 can be made in a similar manner to its bias approximation. Because

$$Var\left\{\hat{f}_j(x)\right\} = \frac{1}{n}\left[\int K^2_{j(x,b)}(u) f(u) du - E^2\left\{f\left(\theta_{j(x,b)}\right)\right\}\right]$$
$$= \frac{1}{n}\int K^2_{j(x,b)}(u) f(u) du + O\left(n^{-1}\right),$$

we may focus on approximating the integral. The integral can be rewritten as

$$\int K^2_{j(x,b)}(u) f(u) du := A_{j(x,b)} E\left\{f\left(\vartheta_{j(x,b)}\right)\right\},$$

where $A_{j(x,b)} := \int K^2_{j(x,b)}(u) du$ is a kernel-specific function of $(x, b)$ (which usually involves the gamma function) and $\vartheta_{j(x,b)}$ is the random variable that obeys the distribution with the pdf $K^2_{j(x,b)}(\cdot) / A_{j(x,b)}$. The leading variance term can be obtained by making an approximation to $A_{j(x,b)}$ and recognizing that $E\left\{f\left(\vartheta_{j(x,b)}\right)\right\} = f(x) + o(1)$. Typically approximation strategies for $A_{j(x,b)}$ differ, depending on whether the design point $x$ is located away from or in the vicinity of the boundary. As a consequence, different convergence rates can be obtained for these regions.

Variance approximations to $\hat{f}_j(x)$ can be documented, depending on whether the support is on $[0, 1]$ or $\mathbb{R}_+$. In order to describe different asymptotic properties of an asymmetric kernel estimator across positions of the design point $x$, we may rely on the phrases "interior $x$" and "boundary $x$" whenever necessary. The design point $x$ is said to be an interior $x$ if $x \in [0, 1]$ satisfies $x/b \to \infty$ and $(1 - x)/b \to \infty$, or if $x \in \mathbb{R}_+$ satisfies $x/b \to \infty$, as $n \to \infty$. On the other hand, $x$ is said to be a boundary $x$ if $x \in [0, 1]$ satisfies either $x/b \to \kappa$ or $(1 - x)/b \to \kappa$, or if $x \in \mathbb{R}_+$ satisfies $x/b \to \kappa$ for some $\kappa \in (0, \infty)$, as $n \to \infty$.

For notational convenience, let

$$v_j = \begin{cases} \frac{1}{2\sqrt{\pi}} & \text{for } j \in \{B, MB, G, MG\} \\ \frac{1}{2\sqrt{2}} & \text{for } j = W \\ \frac{1}{\sqrt{2\pi}} & \text{for } j = NM \end{cases} \quad \text{and} \quad (2.4)$$

$$g_j(x) = \begin{cases} \frac{1}{\sqrt{x(1-x)}} & \text{for } j \in \{B, MB\} \\ \frac{1}{\sqrt{x}} & \text{for } j \in \{G, MG, W, NM\} \end{cases}.$$

We present variance approximations of density estimators using the B and MB kernels first. For interior $x$, recognizing $v_B = 1/\left(2\sqrt{\pi}\right)$ yields

$$Var\left\{\hat{f}_B(x)\right\} = \frac{1}{nb^{1/2}} v_B g_B(x) f(x) + o\left(n^{-1}b^{-1/2}\right)$$

$$= \frac{1}{nb^{1/2}} \frac{f(x)}{2\sqrt{\pi}\sqrt{x(1-x)}} + o\left(n^{-1}b^{-1/2}\right). \qquad (2.5)$$

The dominant term in $Var\left\{\hat{f}_{MB}(x)\right\}$ for interior $x$ takes the same form as the one in $Var\left\{\hat{f}_B(x)\right\}$. In contrast, for boundary $x$, orders of magnitude in $Var\left\{\hat{f}_B(x)\right\}$ and $Var\left\{\hat{f}_{MB}(x)\right\}$ slow down to $O\left(n^{-1}b^{-1}\right)$. In particular, the approximation to $Var\left\{\hat{f}_B(x)\right\}$ for this case is given by

$$Var\left\{\hat{f}_B(x)\right\} = \frac{1}{nb} \frac{\Gamma(2\kappa+1)}{2^{2\kappa+1}\Gamma^2(\kappa+1)} f(x) + o\left(n^{-1}b^{-1}\right).$$

However, it suffices to obtain the order of magnitude for boundary $x$, because the inferior variance convergence does not affect the global property of the density estimator, as can be seen shortly.

Next, we turn to $Var\left\{\hat{f}_j(x)\right\}$ for $j \in \{G, MG, W, NM\}$. For interior $x$,

$$Var\left\{\hat{f}_j(x)\right\} = \frac{1}{nb^{1/2}} v_j g_j(x) f(x) + o\left(n^{-1}b^{-1/2}\right)$$

$$= \frac{1}{nb^{1/2}} v_j \frac{f(x)}{\sqrt{x}} + o\left(n^{-1}b^{-1/2}\right).$$

For boundary $x$, the order of magnitude in $Var\left\{\hat{f}_j(x)\right\}$ again slows down to $O\left(n^{-1}b^{-1}\right)$. In particular, $Var\left\{\hat{f}_G(x)\right\}$ for this case can be approximated as

$$Var\left\{\hat{f}_G(x)\right\} = \frac{1}{nb} \frac{\Gamma(2\kappa+1)}{2^{2\kappa+1}\Gamma^2(\kappa+1)} f(x) + o\left(n^{-1}b^{-1}\right).$$

For each of other three kernels, the Proof of Theorem 2 in Hirukawa and Sakudo (2015) documents $V_B(2, \kappa)$ given in (2.6) below.

*Remark 2.5* $Var\left\{\hat{f}_j(x)\right\}$ for $j \in \{MG, W, NM\}$ can be also obtained via $Var\left\{\hat{f}_{GG}(x)\right\}$. Conditions 1.4 and 1.5 serve as high-level assumptions, and these conditions jointly imply the leading variance term. Approximations to $Var\left\{\hat{f}_{GG}(x)\right\}$ can be summarized as

$$Var\left\{\hat{f}_{GG}(x)\right\} = \begin{cases} \frac{1}{nb^{1/2}} V_I(2) \frac{f(x)}{\sqrt{x}} + o\left(n^{-1}b^{-1/2}\right) & \text{if } x/b \to \infty \\ \frac{1}{nb} V_B(2, \kappa) f(x) + o\left(n^{-1}b^{-1}\right) & \text{if } x/b \to \kappa \end{cases}, \qquad (2.6)$$

where $V_I(2)$ corresponds to $v_j$ in (2.4).

*Remark 2.6* As pointed out by Chen (2000) and Scaillet (2004), a unique feature of the density estimator (2.1) with support on $\mathbb{R}_+$ is that the variance coefficient is inversely proportional to $x^{1/2}$ and thus decreases as $x$ increases. To see this, consider $\hat{f}_G(x)$ for simplicity. If $\theta_{G(x,b)} \overset{d}{=} G(x/b + 1, b)$, then $Var\left(\theta_{G(x,b)}\right) = xb + b^2$. This indicates that the G kernel spreads out as $x$ grows. The varying shape induces the G kernel to possess an adaptive smoothing property similar to the variable kernel method with an effective bandwidth of $xb$ employed, although a single smoothing parameter $b$ is used everywhere. In other words, the kernel can collect more data points (or increase its effective sample size) to smooth in the areas with fewer observations. This property is particularly advantageous to estimating the distributions that have a long tail with sparse data, such as those of the economic and financial variables mentioned in Sect. 1.3.1.

*Remark 2.7* We can compare variance approximations of $\hat{f}_j(x)$ and $\hat{f}_S(x)$. It is well known that the variance of $\hat{f}_S(x)$ can be approximated as

$$Var\left\{\hat{f}_S(x)\right\} = \frac{1}{nh} R(K) f(x) + o\left(n^{-1}h^{-1}\right),$$

where $R(K) = \int_{-\infty}^{\infty} K^2(u)\, du$ is called the roughness of the kernel $K$. Observe that $R(K)$ corresponds to $v_j g_j(x)$. The property that shapes of an asymmetric kernel vary with $x$ makes its "roughness" no longer constant.

## 2.2 Local Properties

Consider the mean squared error ("MSE") of $\hat{f}_j(x)$

$$MSE\left\{\hat{f}_j(x)\right\} = E\left[\left\{\hat{f}_j(x) - f(x)\right\}^2\right] = Bias^2\left\{\hat{f}_j(x)\right\} + Var\left\{\hat{f}_j(x)\right\}.$$

In particular, the MSE for interior $x$ can be approximated by

$$MSE\left\{\hat{f}_j(x)\right\} = \left\{\mathcal{B}_{1,j}(x, f)\right\}^2 b^2 + \frac{1}{nb^{1/2}} v_j g_j(x) f(x) + o\left(b^2 + n^{-1}b^{-1/2}\right).$$

This approximation demonstrates the bias-variance trade-off in $\hat{f}_j(x)$. Observe that the squared bias and variance terms are monotonously increasing and decreasing in $b$, respectively. The optimal smoothing parameter $b_j^*$ that minimizes the dominant two terms in the MSE is

$$b_j^* = \left[\frac{v_j g_j(x) f(x)}{4\left\{\mathcal{B}_{1,j}(x, f)\right\}^2}\right]^{2/5} n^{-2/5}. \tag{2.7}$$

Notice that the order of magnitude in the MSE-optimal smoothing parameter $b_j^*$ is $O\left(n^{-2/5}\right) = O\left(h^{*2}\right)$, where $h^*$ is the MSE-optimal bandwidth for density estimators using nonnegative symmetric kernels. Therefore, when best implemented, the approximation to the MSE reduces to

$$MSE^*\left\{\hat{f}_j\left(x\right)\right\} \sim \frac{5}{4^{5/4}} v_j^{4/5} \left\{\mathcal{B}_{1,j}\left(x,f\right)\right\}^{2/5} \left\{g_j\left(x\right) f\left(x\right)\right\}^{4/5} n^{-4/5}. \quad (2.8)$$

The optimal MSE of $\hat{f}_j\left(x\right)$ for interior $x$ becomes $O\left(n^{-4/5}\right)$, which is also the optimal convergence rate in the MSE of nonnegative symmetric kernel density estimators.

On the other hand, for boundary $x$, $MSE\left\{\hat{f}_j\left(x\right)\right\} = O\left(b^2 + n^{-1}b^{-1}\right)$, and the bias-variance trade-off is again observed. The MSE-optimal smoothing parameter is $b_j^\dagger = O\left(n^{-1/3}\right)$, which yields the optimal MSE of $O\left(n^{-2/3}\right)$. However, it can be soon demonstrated that influence of the slow convergence in the boundary region to the global property of $\hat{f}_j\left(x\right)$ is negligible.

*Remark 2.8* The order of magnitude in $MSE\left\{\hat{f}_j\left(x\right)\right\}$ for interior $x$ is $O\left(b^2 + n^{-1}b^{-1/2}\right)$, which can be rewritten as $O\left(h^4 + n^{-1}h^{-1}\right)$ by defining $h := b^{1/2}$. Some authors pay attention to the fact that $O\left(h^4 + n^{-1}h^{-1}\right)$ is also the order of magnitude in $MSE\left\{\hat{f}_S\left(x\right)\right\}$, the MSE of a nonnegative symmetric kernel density estimator using the bandwidth $h$. Accordingly, they prefer to denote the smoothing parameter by $h^2$ rather than $b$ (e.g., Jones and Henderson 2007).

*Remark 2.9* It is of interest that combining (2.5) with (2.8) leads to

$$MSE^*\left\{\hat{f}_{MB}\left(x\right)\right\} \sim \frac{5}{4} \left(\frac{1}{4\pi}\right)^{2/5} \left\{f^{(2)}\left(x\right)\right\}^{2/5} \left\{f\left(x\right)\right\}^{4/5} n^{-4/5}$$

for interior $x$. The right-hand side coincides with the optimal MSE of the density estimator using the Gaussian kernel $K\left(u\right) = \exp\left(-u^2/2\right)/\sqrt{2\pi}$.

*Remark 2.10* Because $\mathcal{B}_{1,GG}\left(x,f\right) = \left(C_5/2\right) x f^{(2)}\left(x\right)$ for $x \geq C_1 b$, the bias of $\hat{f}_{GG}\left(x\right)$ may become large (in absolute value) as $x$ moves away from the origin. However, the large bias for a large $x$ is compensated (or balanced) by the shrinking variance. Substituting (2.3) and (2.6) into (2.8) yields the optimal MSE for $\hat{f}_{GG}\left(x\right)$ for interior $x$ as

$$MSE^*\left\{\hat{f}_{GG}\left(x\right)\right\} \sim \frac{5}{4} C_5^{2/5} \left\{V_I\left(2\right)\right\}^{4/5} \left\{f^{(2)}\left(x\right)\right\}^{2/5} \left\{f\left(x\right)\right\}^{4/5} n^{-4/5}.$$

Observe that $MSE^* \left\{ \hat{f}_{GG}(x) \right\}$ depends only on $f(x)$ and not on $x$ itself, which suggests that the large bias for a large $x$ is indeed balanced by the shrinking variance. Scaillet (2004) also reports a similar result between bias and variance of the density estimator smoothed by the inverse Gaussian or reciprocal inverse Gaussian kernel.

*Remark 2.11* It is also possible to look into the local properties of density estimators using three special cases of the GG kernels individually. It follows from (2.4) and (2.7) that the MSE-optimal smoothing parameters of the MG, W, and NM density estimators for interior $x$ are

$$b_{MG}^* = \left( \frac{1}{2\sqrt{\pi}} \right)^{2/5} \left[ \frac{\{f(x)\}^{2/5}}{x \left\{ f^{(2)}(x) \right\}^{4/5}} \right] n^{-2/5},$$

$$b_W^* = \left\{ \frac{(12/\pi^2)^2}{2^{3/2}} \right\}^{2/5} \left[ \frac{\{f(x)\}^{2/5}}{x \left\{ f^{(2)}(x) \right\}^{4/5}} \right] n^{-2/5}, \text{ and}$$

$$b_{NM}^* = \left( \frac{2^{3/2}}{\sqrt{\pi}} \right)^{2/5} \left[ \frac{\{f(x)\}^{2/5}}{x \left\{ f^{(2)}(x) \right\}^{4/5}} \right] n^{-2/5}.$$

It can be immediately found that $b_{NM}^* = 2 b_{MG}^*$ holds. By (2.8), the optimal MSEs of $\hat{f}_{MG}(x)$ and $\hat{f}_{NM}(x)$ for interior $x$ have the relation

$$MSE^* \left\{ \hat{f}_{MG}(x) \right\} \sim MSE^* \left\{ \hat{f}_{NM}(x) \right\}$$

$$\sim \frac{5}{4} \left( \frac{1}{4\pi} \right)^{2/5} \left\{ f^{(2)}(x) \right\}^{2/5} \{ f(x) \}^{4/5} n^{-4/5}. \qquad (2.9)$$

The right-hand side is again the optimal MSE of the Gaussian kernel density estimator. In sum, when best implemented, the MG and NM density estimators become first-order asymptotically equivalent. Combining with Remark 2.9, we can see that, the MG and NM kernels on $\mathbb{R}_+$, as well as the MB kernel on $[0, 1]$, are in a sense equivalent to the Gaussian kernel on $\mathbb{R}$. In contrast, when best implemented, the MSE of $\hat{f}_W(x)$ for interior $x$ can be approximated by

$$MSE^* \left\{ \hat{f}_W(x) \right\} \sim \frac{5}{4} \left( \frac{\pi^2}{96} \right)^{2/5} \left\{ f^{(2)}(x) \right\}^{2/5} \{ f(x) \}^{4/5} n^{-4/5}. \qquad (2.10)$$

Comparing the factors of (2.9) and (2.10) reveals that $(5/4) \{ 1/(4\pi) \}^{2/5} \approx 0.454178\ldots$ and $(5/4) \left( \pi^2/96 \right)^{2/5} \approx 0.503178\ldots$ Therefore, $\hat{f}_W(x)$ is shown to be slightly inefficient than $\hat{f}_{MG}(x)$ and $\hat{f}_{NM}(x)$ in the best-case scenario.

## 2.3  Global Properties

Although the variance convergence of $\hat{f}_j(x)$ for boundary $x$ slows down, the inferior convergence rate does not affect its global property. Applying the trimming argument in Chen (1999, p. 136, 2000, p. 476) approximates the mean integrated squared error ("MISE") of $\hat{f}_j(x)$ as

$$
MISE\left\{\hat{f}_j(x)\right\}
$$

$$
= \int E\left[\left\{\hat{f}_j(x) - f(x)\right\}^2\right] dx
$$

$$
= b^2 \int \left\{\mathcal{B}_{1,j}(x, f)\right\}^2 dx + \frac{v_j}{nb^{1/2}} \int g_j(x) f(x) dx + o\left(b^2 + n^{-1}b^{-1/2}\right),
$$

provided that both $\int \left\{\mathcal{B}_{1,j}(x, f)\right\}^2 dx$ and $\int g_j(x) f(x) dx$ are finite, where $\mathcal{B}_{1,j}$ for $j = MB$ and for $j = GG$ refers to the one for $x \in [2b, 1 - 2b]$ and for $x \geq C_1 b$, respectively. The smoothing parameter value that minimizes two dominant terms on the right-hand side is

$$
b_j^{**} = \left(\frac{v_j}{4}\right)^{2/5} \left[\frac{\int g_j(x) f(x) dx}{\int \left\{\mathcal{B}_{1,j}(x, f)\right\}^2 dx}\right]^{2/5} n^{-2/5}.
$$

Therefore, when best implemented, the approximation to the MISE reduces to

$$
MISE^{**}\left\{\hat{f}_j(x)\right\}
$$

$$
\sim \frac{5}{4^{5/4}} v_j^{4/5} \left[\int \left\{\mathcal{B}_{1,j}(x, f)\right\}^2 dx\right]^{1/5} \left\{\int g_j(x) f(x) dx\right\}^{4/5} n^{-4/5}.
$$

In particular, for $\hat{f}_{GG}(x)$, it holds that

$$
MISE^{**}\left\{\hat{f}_{GG}(x)\right\}
$$

$$
\sim \frac{5}{4} C_5^{2/5} \left\{V_I(2)\right\}^{4/5} \left[\int_0^\infty x^2 \left\{f^{(2)}(x)\right\}^2 dx\right]^{1/5} \left\{\int_0^\infty \frac{f(x)}{\sqrt{x}} dx\right\}^{4/5} n^{-4/5}.
$$

Note that $O\left(n^{-4/5}\right)$ is the optimal convergence rate of the MISE within the class of nonnegative kernel estimators in Stone's (1980) sense.

*Remark 2.12* The requirements for finiteness of $\int \left\{\mathcal{B}_{1,j}(x, f)\right\}^2 dx$ and $\int g_j(x) f(x) dx$ depend on kernels. While both $\int \left\{f^{(1)}(x)\right\}^2 dx < \infty$ and $\int x^2 \left\{f^{(2)}(x)\right\}^2 dx < \infty$ imply $\int \left\{\mathcal{B}_{1,j}(x, f)\right\}^2 dx < \infty$ for $j \in \{B, G\}$, the second condition alone establishes $\int \left\{\mathcal{B}_{1,j}(x, f)\right\}^2 dx < \infty$ for $j \in \{MB, GG\}$. Furthermore, for $j \in \{G, GG\}$, $\int_0^\infty g_j(x) f(x) dx = \int_0^\infty \{f(x)/\sqrt{x}\} dx < \infty$ holds as long as $f(x) \propto x^{-q}$ as $x \downarrow 0$ for some $q < 1/2$. Therefore, finiteness of this integral is ensured if the underlying pdf at the origin is zero, takes a positive constant, or even diverges at a certain rate. In contrast, dominant variance coefficients of some other asymmetric kernel density estimators are proportional to $f(x)/x$ or $f(x)/x^{3/2}$. As a result, integrated variances of these estimators are well defined only if $f(0) = 0$. Igarashi and Kakizawa (2017) are concerned with this problem and propose a modification of such kernels. Likewise, for $j \in \{B, MB\}$, $\int_0^1 g_j(x) f(x) dx = \int_0^1 \{f(x)/\sqrt{x(1-x)}\} dx < \infty$ is the case if $f(x) \propto x^{-q}$ as $x \downarrow 0$ and $f(x) \propto (1-x)^{-q}$ as $x \uparrow 1$ for some $q < 1/2$.

**Remark 2.13** Igarashi and Kakizawa (2014) point out that reparameterizing the exponent in the MG kernel can make the MISE of the resulting density estimator even smaller. For two constants $c$ and $c'$ that satisfy $c + 2 > c' \geq 1$, let

$$\rho_{b,c,c'}(x) := \begin{cases} x/b + c & \text{for } x \geq 2b \\ (c - c' + 2)\{x/(2b)\}^{2/(c-c'+2)} + c' & \text{for } x \in [0, 2b) \end{cases}.$$

Also define the pdf of $G\left(\rho_{b,c,c'}(x), b\right)$ as a new class of the G kernel, which is labeled the further modified gamma ("FMG") kernel. Observe that the G and MG kernels are special cases of the FMG kernel with $(c, c') = (1, 1)$ and $(c, c') = (0, 1)$, respectively. It can be demonstrated that the dominant term of the optimal MISE of the density estimator using the FMG kernel is

$$MISE^{**}\left\{\hat{f}_{FMG}(x)\right\}$$
$$\sim \frac{5}{4^{4/5}}\left(\frac{1}{4\pi}\right)^{2/5}\left[\int_0^\infty \{\mathcal{B}_{1,FMG}(x, f)\}^2 dx\right]^{1/5}\left\{\int_0^\infty \frac{f(x)}{\sqrt{x}} dx\right\}^{4/5} n^{-4/5},$$

where

$$\mathcal{B}_{1,FMG}(x, f) = \mathcal{B}_{1,FMG}(x, f; c) := cf^{(1)}(x) + \frac{1}{2}xf^{(2)}(x)$$

depends only on $c$. As in Proposition 3 of Igarashi and Kakizawa (2014), if $x\left\{f^{(1)}(x)\right\}^2 \to 0$ as $x \to \infty$, then using integral by parts yields

$$\int_0^\infty \{\mathcal{B}_{1,FMG}(x, f; c)\}^2 dx$$
$$= c\left(c - \frac{1}{2}\right)\int_0^\infty \{f^{(1)}(x)\}^2 dx + \int_0^\infty \left\{\frac{x}{2}f^{(2)}(x)\right\}^2 dx.$$

The right-hand side is minimized at $c = 1/4$, and the minimum value is

$$
\int_0^\infty \left\{ \mathcal{B}_{1,FMG}\left(x, f; \frac{1}{4}\right) \right\}^2 dx
$$
$$
= \int_0^\infty \left\{ \frac{x}{2} f^{(2)}(x) \right\}^2 dx - \int_0^\infty \left\{ \frac{1}{4} f^{(1)}(x) \right\}^2 dx
$$
$$
< \int_0^\infty \left\{ \frac{x}{2} f^{(2)}(x) \right\}^2 dx \left( = \int_0^\infty \left\{ \mathcal{B}_{1,MG}(x, f) \right\}^2 dx \right).
$$

Putting $c' = 1$ for simplicity, Igarashi and Kakizawa (2014) conclude that $\hat{f}_{FMG}$ with $(c, c') = (1/4, 1)$ is preferred to $\hat{f}_{MG}$ in terms of the MISE.

## 2.4   Other Convergence Results

While our focus so far has been on the mean square convergence, different types of convergence results are also available. Bouezmarni and Rolin (2003) and Bouezmarni and Scaillet (2005) show weak $L_1$ consistency of the B and G density estimators, respectively. Bouezmarni and Rolin (2003) and Bouezmarni and Scaillet (2005) also demonstrate uniform weak consistency of these estimators. Furthermore, Bouezmarni and Rolin (2003) and Bouezmarni and Rombouts (2010a) establish their uniform strong consistency. These results are summarized as the theorem below.

**Theorem 2.1**  **(i)** (Bouezmarni and Rolin 2003, Theorem 3 and Remark 2)
*If $f(\cdot)$ has support on $[0, 1]$ and is continuous and bounded on $[0, 1]$, then*

$$
\sup_{x \in [0,1]} \left| \hat{f}_B(x) - f(x) \right| \begin{cases} \xrightarrow{p} 0 \ if \ b + \left(nb^2\right)^{-1} \to 0 \\ \xrightarrow{a.s.} 0 \ if \ b + \log n / \left(nb^2\right) \to 0 \end{cases}.
$$

**(ii)** (Bouezmarni and Scaillet 2005, Theorem 3.1; Bouezmarni and Rombouts 2010a, Theorem 2)
*If $f(\cdot)$ has support on $\mathbb{R}_+$ and is continuous and bounded on a compact interval $I \subset \mathbb{R}_+$, then*

$$
\sup_{x \in I} \left| \hat{f}_G(x) - f(x) \right| \begin{cases} \xrightarrow{p} 0 \ if \ b + \left(nb^2\right)^{-1} \to 0 \\ \xrightarrow{a.s.} 0 \ if \ b + \log n / \left(nb^2\right) \to 0 \end{cases}.
$$

## 2.5  Properties of Density Estimators at the Boundary

### 2.5.1  Bias and Variance of Density Estimators at the Boundary

This section explores properties of the density estimator (2.1) at the boundary. We start from approximating the bias and variance of $\hat{f}_j(x)$ at the boundary under the assumption that the true density is finite at the boundary. This analysis is a natural extension of the one in Theorem 3.2 of Shi and Song (2016). The results are documented separately, depending on whether the support is on [0, 1] or $\mathbb{R}_+$.

At each boundary (i.e., $x = 0, 1$) biases and variances of $\hat{f}_B(x)$ and $\hat{f}_{MB}(x)$ are first-order asymptotically equivalent. Specifically, for $j \in \{B, MB\}$,

$$Bias\left\{\hat{f}_j(0)\right\} = f^{(1)}(0)b + o(b), \tag{2.11}$$

$$Var\left\{\hat{f}_j(0)\right\} = \begin{cases} \frac{1}{nb}\frac{f(0)}{2} + o\left(n^{-1}b^{-1}\right) & \text{if } f(0) > 0 \\ \frac{1}{n}\frac{f^{(1)}(0)}{4} + o\left(n^{-1}\right) & \text{if } f(0) = 0 \text{ and } f^{(1)}(0) > 0 \end{cases}, \tag{2.12}$$

$$Bias\left\{\hat{f}_j(1)\right\} = -f^{(1)}(1)b + o(b), \text{ and}$$

$$Var\left\{\hat{f}_j(1)\right\} = \begin{cases} \frac{1}{nb}\frac{f(1)}{2} + o\left(n^{-1}b^{-1}\right) & \text{if } f(1) > 0 \\ -\frac{1}{n}\frac{f^{(1)}(1)}{4} + o\left(n^{-1}\right) & \text{if } f(1) = 0 \text{ and } f^{(1)}(1) < 0 \end{cases}.$$

Biases and variances of $\hat{f}_j(0)$ for $j \in \{G, MG, W\}$ admit the same first-order expansions as in (2.11) and (2.12), respectively. While $\hat{f}_{NM}(0)$ also has the same first-order bias expansion, its variance expansion slightly differs. It is

$$Var\left\{\hat{f}_{NM}(0)\right\} = \begin{cases} \frac{1}{nb}\frac{\sqrt{2}}{\pi}f(0) + o\left(n^{-1}b^{-1}\right) & \text{if } f(0) > 0 \\ \frac{1}{n}\frac{f^{(1)}(0)}{\pi} + o\left(n^{-1}\right) & \text{if } f(0) = 0 \text{ and } f^{(1)}(0) > 0 \end{cases}.$$

### 2.5.2  Consistency of Density Estimators for Unbounded Densities at the Origin

So far we have delivered asymptotic results of asymmetric kernel density estimators under the assumption that $f$ is bounded at the boundary. In the previous section, the estimators are shown to be indeed consistent at the boundary under this assumption. In reality, the assumption may be violated. For example, a clustering of observations near the boundary, or a pole in the density at the origin can be frequently observed in distributions of intraday trading volumes (e.g., Malec and Schienle 2014) and spectral densities of long memory processes (e.g., Bouezmarni and Van Bellegem 2011).

We may still employ the B, G, and GG kernels to estimate such unbounded densities consistently. Several authors demonstrate weak consistency and the relative convergence of the density estimators for unbounded densities. The next two theorems document these results.

**Theorem 2.2** (Bouezmarni and Scaillet 2005, Theorem 5.1; Bouezmarni and Van Bellegem 2011, Proposition 3.3; Hirukawa and Sakudo 2015, Theorem 5)

If $f(x)$ is unbounded at $x = 0$ and $b + \left(nb^2\right)^{-1} \to 0$ as $n \to \infty$, then $\hat{f}_j(0) \xrightarrow{p} \infty$ for $j \in \{B, G, GG\}$.

**Theorem 2.3** (Bouezmarni and Scaillet 2005, Theorem 5.3; Bouezmarni and Van Bellegem 2011, Proposition 3.4; Hirukawa and Sakudo 2015, Theorem 6)

If $f(x)$ is unbounded at $x = 0$ and continuously differentiable in the neighborhood of the origin and $b + \left\{nb^2 f(x)\right\}^{-1} \to 0$ as $n \to \infty$ and $x \to 0$, then

$$\left| \frac{\hat{f}_j(x) - f(x)}{f(x)} \right| \xrightarrow{p} 0$$

for $j \in \{B, G, GG\}$ as $x \to 0$.

*Remark 2.14* Not all asymmetric kernels share the appealing properties described in these theorems. Igarashi and Kakizawa (2014, Sect. 2.2) argue that density estimators smoothed by some asymmetric kernels generate nonnegligible bias at or near the boundary by construction. Specifically, both the Birnbaum–Saunders kernel by Jin and Kawczak (2003) and the inverse Gaussian kernel by Scaillet (2004) always yield zero density estimates at the origin even when the truth is $f(0) > 0$. Moreover, the reciprocal inverse Gaussian kernel by Scaillet (2004) is shown to have downward bias in the vicinity of the origin.

## 2.6 Further Topics

### 2.6.1 Density Estimation Using Weakly Dependent Observations

Often we are interested in estimating the marginal density from nonnegative time-series data. Examples include estimation problems of (i) the distribution of important financial variables such as short-term interest rates or trading volumes and (ii) the baseline hazard in financial duration analysis.

To explore convergence properties of asymmetric kernel density estimators using dependent observations, we must impose some regularity condition on their dependent structure, typically as a notion of mixing. Here, we exclusively focus on $\alpha$-mixing (or strongly mixing) processes by Rosenblatt (1956). For reference, an

$\alpha$-mixing process is formally defined as follows. For $-\infty \leq J \leq L \leq \infty$ denote the $\sigma$-algebra generated from the stationary random variables $\{X_t\}_{t=J}^{L}$ by

$$\mathcal{F}_J^L := \sigma \{X_t, \ J \leq t \leq L \ (t \in \mathbb{Z})\}$$

and define

$$\alpha(\tau) := \sup_{j \in \mathbb{Z}} \sup_{A \in \mathcal{F}_{-\infty}^j, B \in \mathcal{F}_{j+\tau}^\infty} |\Pr(A \cap B) - \Pr(A)\Pr(B)|.$$

The stationary process $\{X_t\}_{t=-\infty}^{\infty}$ is said to be $\alpha$-mixing if $\alpha(\tau) \to 0$ as $\tau \to \infty$. As an alternative condition, $\beta$-mixing (or absolute regularity) is often considered in economics and finance (e.g., Carrasco and Chen 2002; Chen et al. 2010). Since $\beta$-mixing implies $\alpha$-mixing, assuming the latter is general enough to cover many important applications in economics and finance.

As with density estimators using standard symmetric kernels, the leading bias and variance terms for asymmetric kernel density estimators remain unchanged even when positive weakly dependent observations are used. Intuitively, even if the observations chosen for a local average in the neighborhood of a design point have serial dependence, they are not necessarily close to each other in time as long as their dependence is weak. As a result, the dependent observations are likely to behave as if they were independent.

Bouezmarni and Rombouts (2010b, Proposition 1) first extend density estimation using asymmetric kernels in this direction. They exclusively consider the MG kernel and show that the (first-order) bias and variance approximations are still valid for $\alpha$-mixing processes with an exponentially decaying mixing coefficient $\alpha(\tau)$ such that $\alpha(\tau) \leq C\rho^\tau$ for some $C \in (0, \infty)$ and $\rho \in (0, 1)$. The results hold true even when the decay rate of $\alpha(\tau)$ is relaxed to a polynomial one. Under $\alpha$-mixing of size $-(2r-2)/(r-2)$ for some $r > 2$, i.e., $\alpha(\tau) \leq C\tau^{-q}$ for some $q > (2-2/r)/(1-2/r)$, Hirukawa and Sakudo (2015, Theorems 4–6) demonstrate that not only the bias and variance approximations but also two theorems on estimating unbounded densities at the origin (i.e., Theorems 2.2 and 2.3) hold for the entire family of the GG kernels including the MG kernel.

## 2.6.2  Normalization

An asymmetric kernel $K_{x,b}(\cdot)$ cannot be expressed in the form of $(1/b) K \{(\cdot - x)/b\}$, or roles of the design and data points are nonexchangeable. The Gaussian-copula kernel by Jones and Henderson (2007) is an exception in that the roles are exchangeable inside this kernel by construction. The lack of exchangeability incurs a cost. Asymmetric kernel density estimators are not bona fide in general, in the sense that they are not integrated to unity in finite samples. To be more precise, $\int \hat{f}_j(x) \, dx = 1 + O(b)$, provided that $\mathcal{B}_{1,j}(x, f)$ is absolutely integrable.

The normalization problem is common across most of asymmetric kernels (including those under consideration in this book). Jones and Henderson (2007) even refer to (2.1) as a kernel-type density estimator, where the phrase "kernel-type" signifies the fact that they lack a basic property every symmetric kernel possesses. Accordingly, it could be also more appropriate to classify asymmetric kernels not as kernels but as kernel-type weighting functions. Nonetheless, following the convention in the literature, we simply express them as kernels throughout.

Here are a few remarks on this matter. First, the normalization problem is not uncommon in density estimation smoothed by symmetric kernels, actually. For example, sometimes symmetric kernels are employed to estimate a density with support on $\mathbb{R}_+$. The resulting density estimates are not integrated to one over $\mathbb{R}_+$. As will be revisited in Chap. 3, bias-corrected density estimators using symmetric kernels are not bona fide, either; see, for instance, Sect. 2.2 of Jones et al. (1995). Second, there are several attempts to make asymmetric kernel density estimators bona fide. For densities with support on [0, 1], Gouriéroux and Monfort (2006) propose two methods of renormalizing the B kernel density estimator in the context of density estimation for recovery rate distributions. Two renormalized density estimators, namely the "macro-" and "micro-beta" estimators, are defined as

$$\hat{f}_B^R (x) := \frac{\hat{f}_B (x)}{\int_0^1 \hat{f}_B (x)\, dx} \text{ and } \hat{f}_B^r (x) := \frac{1}{n} \sum_{i=1}^n \frac{K_{B(x,b)} (X_i)}{\int_0^1 K_{B(x,b)} (X_i)\, dx},$$

respectively. Observe that in the former renormalization is made after the initial density estimate is obtained, whereas in the latter the B kernel itself is renormalized before density estimation. Jones and Henderson (2007) also propose the Gaussian-copula kernel, in which roles of the design and data points are exchangeable as in symmetric kernels. Moreover, for densities with support on $\mathbb{R}_+$, Jeon and Kim (2013) study the G kernel with roles of the design and data points reversed.

### 2.6.3  Extension to Multivariate Density Estimation

When multivariate bounded data are available, we may be interested in estimating their joint density, as well as individual marginal densities. There are only a few studies on multivariate extension of asymmetric kernel density estimation, to the best of our knowledge. Examples include Bouezmarni and Rombouts (2010a) and Funke and Kawka (2015). In principle, there are two approaches for estimating the joint density of multivariate bounded random variables by asymmetric kernels. One way is to apply the product kernel method, and the other is to employ multivariate extensions of beta, gamma, and generalized gamma densities as kernels.

The former is a natural and straightforward strategy, and both Bouezmarni and Rombouts (2010a) and Funke and Kawka (2015) adopt it. A benefit of this strategy is that asymptotic results on the joint density estimators can be obtained easily and are similar to those for one-dimensional cases. In particular, Funke and Kawka (2015) investigate multivariate bias-corrected density estimation, and convergence properties of their estimators will be revisited in Chap. 3.

On the other hand, the latter may not be practical. Density functions for multivariate versions of beta, gamma, and generalized gamma distributions take complicated forms. Even though they can be used as kernels, it appears to be quite challenging to explore convergence properties of the joint density estimators smoothed by such kernels.

## 2.7 Smoothing Parameter Selection

Choosing the smoothing parameter $b$ is an important practical issue. If the value of $b$ is appropriately chosen, then it can help to yield a density estimate that is close to the truth; however, a poorly selected smoothing parameter is likely to distort the quality of the density estimate severely.

There are several bandwidth selection methods available for symmetric kernel density estimation; see Jones et al. (1996) for a brief survey. In contrast, it seems that the problem of choosing a smoothing parameter for asymmetric kernel density estimation still stays at the stage of what Jones et al. (1996) categorize as "first generation methods." Below two main approaches, namely plug-in (or rule-of-thumb) and cross-validation methods, are discussed in order.

It is worth emphasizing that we exclusively consider the problem of choosing a single, global smoothing parameter that can be used everywhere. It could be also possible to vary smoothing parameter values across different design points. However, we do not pursue this idea because each asymmetric kernel can change the amount of smoothing by varying its shapes under a single smoothing parameter; see Hagmann and Scaillet (2007, p. 229) for a discussion.

### 2.7.1 Plug-In Methods

A plug-in method is built on the minimizer of two dominant terms in the MISE of $\hat{f}_j(x)$. Because both terms include an unknown quantity, they are replaced by those implied by a parametric (or reference) density. In this sense, this approach is very close to Silverman's (1986) rule-of-thumb bandwidth, where a normal density is chosen as the reference. Examples of plug-in smoothing parameters can be found in Scaillet (2004), Hirukawa (2010) and Hirukawa and Sakudo (2015).

It is straightforward to derive the plug-in smoothing parameters for the MB and GG density estimators. For the MB density estimator, Hirukawa (2010) defines the plug-in smoothing parameter $\hat{b}_{MB}$ as

$$\hat{b}_{MB} = \arg\min_b \left[ \frac{b^2}{4} \int_0^1 x^2 (1-x)^2 \left\{ f_\theta^{(2)}(x) \right\}^2 v(x)\, dx \right.$$

$$\left. + \frac{n^{-1} b^{-1/2}}{2\sqrt{\pi}} \int_0^1 \frac{f_\theta(x)}{\sqrt{x(1-x)}} v(x)\, dx \right].$$

While the objective function is in principle based on two dominant terms of $MISE\left\{ \hat{f}_{MB}(x) \right\}$, a couple of modifications are made. First, the unknown true density $f$ is replaced by a reference $f_\theta$ with some parameter $\theta$. Second, a weight function $v$ is introduced to ensure finiteness of two integrals. Specifically, the pdf of $Beta\,(\alpha, \beta)$ is chosen as $f_\theta$, and the weight function is specified as $v(x) = x^3 (1-x)^3$. It follows that $\hat{b}_{MB}$ has an explicit form of

$$\hat{b}_{MB} = \left\{ \frac{B(\alpha, \beta)\, B(\alpha + 5/2, \beta + 5/2)}{2\sqrt{\pi} C_{\alpha, \beta}} \right\}^{2/5} n^{-2/5},$$

where

$$C_{\alpha, \beta} = \frac{1}{\Gamma(2\alpha + 2\beta + 4)} \left\{ (\alpha - 1)^2 (\alpha - 2)^2 \Gamma(2\alpha)\, \Gamma(2\beta + 4) \right.$$

$$- 4(\alpha - 1)^2 (\alpha - 2)(\beta - 1)\, \Gamma(2\alpha + 1)\, \Gamma(2\beta + 3)$$

$$+ 2(\alpha - 1)(\beta - 1)(3\alpha\beta - 4\alpha - 4\beta + 6)\, \Gamma(2\alpha + 2)\, \Gamma(2\beta + 2)$$

$$- 4(\alpha - 1)(\beta - 1)^2 (\beta - 2)\, \Gamma(2\alpha + 3)\, \Gamma(2\beta + 1)$$

$$\left. + (\beta - 1)^2 (\beta - 2)^2 \Gamma(2\alpha + 4)\, \Gamma(2\beta) \right\}.$$

In practice, the parameter $\theta = (\alpha, \beta)$ should be replaced by some consistent estimate via maximum likelihood ("ML") or method of moments.

Moreover, if the true distribution is $U[0, 1]$ so that $\alpha = \beta = 1$, then $\hat{b}_{MB}$ tends to take a large value. Invoke that $\hat{f}_{MB}(x)$ (as well as $\hat{f}_B(x)$) is unbiased for the pdf of $U[0, 1]$ in Remark 2.1. It follows that the smoothing parameter that minimizes the sum of two dominant terms in the MISE is not well defined (even after some modifications are made for both terms). Under this circumstance, it suffices to employ a $O(1)$ smoothing parameter, and a large $\hat{b}_{MB}$ simply reflects "optimality" of oversmoothing.

Hirukawa (2010) does not derive $\hat{b}_B$, yet another smoothing parameter for $\hat{f}_B(x)$, because extra terms involving $f^{(1)}(\cdot)$ in its integrated squared bias cause $\hat{b}_B$ to have many terms involving $(\alpha, \beta)$. It follows that $\hat{b}_B$ tends to be noisy because of estimation errors of the parameters. Monte Carlo simulations in Hirukawa (2010) indicate no adversity in using $\hat{b}_{MB}$ for $\hat{f}_B(x)$.

The plug-in smoothing parameter for the GG density estimator (including the MG, W, and NM density estimators as special cases) due to Hirukawa and Sakudo (2015) can be obtained in a similar manner. It is defined as

$$\hat{b}_{GG} = \arg\min_b \left[ b^2 \left( \frac{C_5^2}{4} \right) \int_0^\infty x^2 \left\{ f_\theta^{(2)}(x) \right\}^2 v(x)\, dx \right.$$
$$\left. + \frac{V_I(2)}{nb^{1/2}} \int_0^\infty \frac{f_\theta(x)}{\sqrt{x}} v(x)\, dx \right],$$

where the pdf of $G(\gamma, \delta)$ is chosen as the reference $f_\theta$ and the weight function is specified as $v(x) = x^3$. As a result, the plug-in smoothing parameter can be simplified to

$$\hat{b}_{GG} = \left\{ \frac{4^{\gamma-1} V_I(2)\, \delta^{5/2} \Gamma(\gamma) \Gamma(\gamma + 5/2)}{C_5^2 C_\gamma \Gamma(2\gamma)} \right\}^{2/5} n^{-2/5},$$

where

$$C_\gamma = \frac{1}{4} (\gamma - 2)^2 (\gamma - 1)^2 - (\gamma - 2)(\gamma - 1)^2 \gamma$$
$$+ \frac{1}{2} (3\gamma - 4)(\gamma - 1)\gamma \left( \gamma + \frac{1}{2} \right) - (\gamma - 1)\gamma \left( \gamma + \frac{1}{2} \right)(\gamma + 1)$$
$$+ \frac{1}{4} \gamma \left( \gamma + \frac{1}{2} \right)(\gamma + 1)\left( \gamma + \frac{3}{2} \right).$$

As before, we must replace the parameter $\theta = (\gamma, \delta)$ with some consistent estimate to implement $\hat{b}_{GG}$.

## 2.7.2 Cross-Validation Methods

Cross-validation ("CV") is another popular method of choosing the smoothing parameter. The idea of CV is to find the value of $b$ that minimizes the integrated squared error (" ISE")

$$ISE\left\{ \hat{f}_j(x) \right\} = \int \left\{ \hat{f}_j(x) - f(x) \right\}^2 dx$$
$$= \int \left\{ \hat{f}_j(x) \right\}^2 dx - 2 \int \hat{f}_j(x) f(x)\, dx + \int \{ f(x) \}^2 dx.$$

Because the last term does not depend on $b$, an estimate of $ISE\left\{ \hat{f}_j(x) \right\} - \int \{ f(x) \}^2 dx$ serves as the CV criterion. Let

$$\hat{f}_{j,-i}(X_i) := \frac{1}{n-1} \sum_{k=1, k \neq i}^n K_{j(X_i, b)}(X_k)$$

be the density estimate using the sample with $X_i$ removed. Then, $\int \hat{f}_j(x) f(x) dx$ can be estimated as

$$\frac{1}{n} \sum_{i=1}^{n} \hat{f}_{j,-i}(X_i) = \frac{1}{n(n-1)} \sum_{i=1}^{n} \sum_{k=1, k \neq i}^{n} K_{j(X_i, b)}(X_k).$$

It follows that the leave-one-out CV criterion becomes

$$CV(b) = \int \left\{ \hat{f}_j(x) \right\}^2 dx - \frac{2}{n(n-1)} \sum_{i=1}^{n} \sum_{k=1, k \neq i}^{n} K_{j(X_i, b)}(X_k),$$

where the integral in the first term can be evaluated numerically. The CV method then selects the smoothing parameter value as $\hat{b}_{CV} = \arg\min_b CV(b)$. Asymptotic optimality of the selector $\hat{b}_{CV}$ for density estimation using the G kernel is demonstrated in Theorem 4 of Bouezmarni and Rombouts (2010a).

However, the leave-one-out CV method may not be appropriate for density estimation using weakly dependent (but possibly persistent) observations. In estimating densities using positive time-series data, Bouezmarni and Rombouts (2010b) replace $CV(b)$ with the $h$-block CV criterion

$$CV_h(b) = \int \left\{ \hat{f}_j(x) \right\}^2 dx - \frac{2}{n} \sum_{i=1}^{n} \hat{f}_{j,-(i-h;i+h)}(X_i),$$

where

$$\hat{f}_{j,-(i-h;i+h)}(X_i) := \frac{1}{n - (2h + 1)} \sum_{k=1, |k-i| > h}^{n} K_{j(X_i, b)}(X_k)$$

is the density estimate using $n - (2h + 1)$ observations obtained by removing $X_i$ itself and $h$ data points on both sides of $X_i$ from the entire $n$ observations; see Chap. VI of Györfi et al. (1989) for more details. The idea behind this procedure is that a diverging block size $h$ (at a certain rate) induces the remaining $n - (2h + 1)$ observations to behave as if they were independent. Obviously, when $h = 0$, $CV_h(b)$ collapses to $CV(b)$ for independent observations. Accordingly, the $h$-block CV method selects the smoothing parameter value as $\hat{b}_{CV_h} = \arg\min_b CV_h(b)$. Notice that for consistency of the procedure, the divergence rate of $h$ must be slower than the sample size $n$, and Bouezmarni and Rombouts (2010b) recommend putting $h = n^{1/4}$. This approach will be revisited in Chap. 4 in the context of nonparametric estimation of scalar diffusion models.

## 2.8   List of Useful Formulae

This section lists several formulae on gamma functions for reference. These are found to be useful for asymptotic expansions of density and regression estimators and test statistics smoothed by asymmetric kernels.

**Moments about Zero of the Beta, Gamma, and Generalized Gamma Distributions**

Let $X \overset{d}{=} Beta\,(p, q)$, $Y \overset{d}{=} G\,(p, q)$ and $Z \overset{d}{=} G\,(p, q, r)$ for $p, q, r > 0$. Then, for $m \in \mathbb{Z}_+$,

$$
\begin{aligned}
E\left(X^m\right) &= \frac{\Gamma\,(p+q)\,\Gamma\,(p+m)}{\Gamma\,(p+q+m)\,\Gamma\,(p)} \\
&= \frac{p\,(p+1)\cdots(p+m-1)}{(p+q)\,(p+q+1)\cdots(p+q+m-1)}, \quad\quad (2.13)
\end{aligned}
$$

$$
\begin{aligned}
E\left(Y^m\right) &= q^m \frac{\Gamma\,(p+m)}{\Gamma\,(p)} \\
&= p\,(p+1)\cdots(p+m-1)\,q^m, \text{ and} \quad\quad (2.14)
\end{aligned}
$$

$$
E\left(Z^m\right) = q^m \frac{\Gamma\,\{(p+m)\,/r\}}{\Gamma\,(p/r)}. \quad\quad (2.15)
$$

**Stirling's Asymptotic Formula**

$$
\Gamma\,(a+1) = \sqrt{2\pi}\,a^{a+1/2}\exp\,(-a)\left\{1 + \frac{1}{12a} + O\left(a^{-2}\right)\right\} \text{ as } a \to \infty. \quad (2.16)
$$

**Series Expansion of the Log Gamma Function**

$$
\log\Gamma\,(1+a) = -\gamma a + \sum_{k=2}^{\infty} \frac{(-1)^k\,\zeta\,(k)}{k}a^k \text{ for } |a| < 1,
$$

where (only in this context)

$$
\gamma = \lim_{n\to\infty}\left(\sum_{k=1}^{n}\frac{1}{k} - \log n\right) = 0.5772156649\ldots
$$

is Euler's constant, and

$$
\zeta\,(k) = \sum_{n=1}^{\infty}\frac{1}{n^k} \text{ for } k > 1
$$

is the Riemann zeta function.

**Legendre Duplication Formula** (Legendre 1809)

$$\Gamma\left(a\right)\Gamma\left(a+\frac{1}{2}\right) = \frac{\sqrt{\pi}}{2^{2a-1}}\Gamma\left(2a\right) \text{ for } a > 0.$$

**Recursive Formula on the Lower Incomplete Gamma Function**

$$\gamma\left(a+1, z\right) = a\gamma\left(a, z\right) - z^{a}\exp\left(-z\right) \text{ for } a, z > 0. \tag{2.17}$$

# References

Bouezmarni, T., and J.-M. Rolin. 2003. Consistency of the beta kernel density function estimator. *Canadian Journal of Statistics* 31: 89–98.

Bouezmarni, T., and J.V.K. Rombouts. 2010a. Nonparametric density estimation for multivariate bounded data. *Journal of Statistical Planning and Inference* 140: 139–152.

Bouezmarni, T., and J.V.K. Rombouts. 2010b. Nonparametric density estimation for positive time series. *Computational Statistics and Data Analysis* 54: 245–261.

Bouezmarni, T., and O. Scaillet. 2005. Consistency of asymmetric kernel density estimators and smoothed histograms with application to income data. *Econometric Theory* 21: 390–412.

Bouezmarni, T., and S. Van Bellegem. 2011. Nonparametric beta kernel estimator for long memory time series, ECORE Discussion Paper 2011/19.

Carrasco, M., and X. Chen. 2002. Mixing and moment properties of various GARCH and stochastic volatility models. *Econometric Theory* 18: 17–39.

Chen, S.X. 1999. Beta kernel estimators for density functions. *Computational Statistics and Data Analysis* 31: 131–145.

Chen, S.X. 2000. Probability density function estimation using gamma kernels. *Annals of the Institute of Statistical Mathematics* 52: 471–480.

Chen, X., L.P. Hansen, and M. Carrasco. 2010. Nonlinearity and temporal dependence. *Journal of Econometrics* 155: 155–169.

Funke, B., and R. Kawka. 2015. Nonparametric density estimation for multivariate bounded data using two non-negative multiplicative bias correction methods. *Computational Statistics and Data Analysis* 92: 148–162.

Gouriéroux, C., and A. Monfort. 2006. (Non)Consistency of the beta kernel estimator for recovery rate distribution, CREST Discussion Paper n° 2006–31.

Györfi, L., W. Härdle, P. Sarda, and P. Vieu. 1989. *Nonparametric Curve Estimation from Time Series.*, Lecture Notes in Statistics Berlin: Springer-Verlag.

Hagmann, M., and O. Scaillet. 2007. Local multiplicative bias correction for asymmetric kernel density estimators. *Journal of Econometrics* 141: 213–249.

Hirukawa, M. 2010. Nonparametric multiplicative bias correction for kernel-type density estimation on the unit interval. *Computational Statistics and Data Analysis* 54: 473–495.

Hirukawa, M., and M. Sakudo. 2015. Family of the generalised gamma kernels: a generator of asymmetric kernels for nonnegative data. *Journal of Nonparametric Statistics* 27: 41–63.

Igarashi, G., and Y. Kakizawa. 2014. Re-formulation of inverse Gaussian, reciprocal inverse Gaussian, and Birnbaum-Saunders kernel estimators. *Statistics and Probability Letters* 84: 235–246.

Igarashi, G., and Y. Kakizawa. 2017. Inverse gamma kernel density estimation for nonnegative data. *Journal of the Korean Statistical Society* 46: 194–207.

Jeon, Y., and J.H.T. Kim. 2013. A gamma kernel density estimation for insurance data. *Insurance: Mathematics and Economics* 53: 569–579.

Jin, X., and J. Kawczak. 2003. Birnbaum-Saunders and lognormal kernel estimators for modelling durations in high frequency financial data. *Annals of Economics and Finance* 4: 103–124.

Jones, M.C., and D.A. Henderson. 2007. Kernel-type density estimation on the unit interval. *Biometrika* 24: 977–984.

Jones, M.C., O. Linton, and J.P. Nielsen. 1995. A simple bias reduction method for density estimation. *Biometrika* 82: 327–338.

Jones, M.C., J.S. Marron, and S.J. Sheather. 1996. A brief survey of bandwidth selection for density estimation. *Journal of the American Statistical Association* 91: 401–407.

Legendre, A.-M. 1809. *Mémoires de la classe des sciences mathématiques et physiques*. Paris: Institut Impérial de France, 477, 485, 490.

Malec, P., and M. Schienle. 2014. Nonparametric kernel density estimation near the boundary. *Computational Statistics and Data Analysis* 72: 57–76.

Rosenblatt, M. 1956. A central limit theorem and a strong mixing condition. *Proceedings of the National Academy of Science of the United States of America* 42: 43–47.

Scaillet, O. 2004. Density estimation using inverse and reciprocal inverse Gaussian kernels. *Journal of Nonparametric Statistics* 16: 217–226.

Shi, J., and W. Song. 2016. Asymptotic results in gamma kernel regression. *Communications in Statistics - Theory and Methods* 45: 3489–3509.

Silverman, B.W. 1986. *Density Estimation for Statistics and Data Analysis*. London: Chapman and Hall.

Stone, C.J. 1980. Optimal rates of convergence for nonparametric estimators. *Annals of Statistics* 8: 1348–1360.

# Chapter 3
# Bias Correction in Density Estimation

We are motivated to estimate the quantity of interest in a less biased manner, and density estimation is not an exception. The bias correction methods discussed in this chapter are natural extensions of what are originally proposed for nonnegative (or second-order) symmetric kernels. Again in this chapter $n$ observations $\{X_i\}_{i=1}^n$ used for density estimation are assumed to be *i.i.d.* random variables drawn from a univariate distribution with a pdf $f$ having support either on $[0, 1]$ or $\mathbb{R}_+$. A variety of (fully) nonparametric and semiparametric bias correction techniques are described. Our main focus is on improvement in the bias property of each bias-corrected estimator. We do not refer to its limiting distribution, although it is not hard to derive it in light of the Proof of Theorem 5.4 in Sect. 5.5.1. Smoothing parameter selection for some bias-corrected estimators is also discussed.

## 3.1 An Overview

Bias correction methods using asymmetric kernels are classified roughly into two approaches, namely nonparametric and semiparametric ones. Below we present an overview of each approach.

### 3.1.1 Nonparametric Bias Correction

Nonparametric methods provide valid inference under a much broader class of structures than those imposed by parametric models. However, there is a price to pay for the robust inference. Nonparametric estimators in general have slower convergence than parametric ones do. The bias induced by kernel smoothing may be substantial even for moderate sample sizes. This motivates us to investigate bias correction (or reduction) for asymmetric kernel density estimation in a fully nonparametric manner.

© The Author(s) 2018
M. Hirukawa, *Asymmetric Kernel Smoothing*, JSS Research Series
in Statistics, https://doi.org/10.1007/978-981-10-5466-2_3

There are a variety of fully nonparametric bias correction techniques for density estimation using nonnegative symmetric kernels; see Jones and Foster (1993) and Jones and Signorini (1997) for reviews. The literature on this approach using asymmetric kernels includes Hirukawa (2010), Hirukawa and Sakudo (2014, 2015), Igarashi and Kakizawa (2015), and Funke and Kawka (2015). Hirukawa (2010) and Hirukawa and Sakudo (2014, 2015) apply two classes of nonparametric multiplicative bias correction methods studied by Terrell and Scott (1980) and Jones et al. (1995) to density estimation with support on [0, 1] and $\mathbb{R}_+$, respectively. The additive bias correction considered by Igarashi and Kakizawa (2015) follows a version of generalized jackknife methods by Jones and Foster (1993), and the approach may be viewed as an attempt to construct fourth-order asymmetric kernels with support on $\mathbb{R}_+$.

Each of the above approaches has a common feature. As seen in Chap. 2, an asymmetric kernel density estimator has a $O(b)$ bias. As demonstrated shortly, under sufficient smoothness of $f$, bias convergence of each bias-corrected estimator can be accelerated from $O(b)$ to $O(b^2)$, whereas the order of magnitude in variance remains unchanged, i.e., it is still $O(n^{-1}b^{-1/2})$ for interior $x$. The accelerated bias convergence leads to a faster convergence rate. Because the MISE of each bias-corrected estimator is $O(b^4 + n^{-1}b^{-1/2})$, it can achieve the convergence rate of $O(n^{-8/9})$ in MISE when best implemented. The rate is faster than $O(n^{-4/5})$, the best-possible convergence rate in MISE within the class of nonnegative kernel estimators. Furthermore, Funke and Kawka (2015) extend the bias correction techniques studied in Hirukawa (2010) and Hirukawa and Sakudo (2014, 2015) to joint density estimation using multivariate bounded data.

### 3.1.2  Semiparametric Bias Correction

Parametric models are still popularly chosen in estimating distributions of economic and financial variables such as incomes and payments to the insured. However, imposition of an imprecise parametric model may lead to an inconsistent density estimate and misleading inference. Then, asymmetric kernels can be employed to reduce the bias induced by inaccuracy in parametric specification. Inevitably the entire bias correction procedure becomes semiparametric in the sense that a nonparametric kernel method helps to decrease the bias of an initial parametric density estimator.

Examples of semiparametric bias correction using asymmetric kernels include Hagmann and Scaillet (2007), Gustafsson et al. (2009), and Hirukawa and Sakudo (2018). Although each method uses asymmetric kernels at the bias correction step after initial parametric density estimation, there is a crucial difference. The local multiplicative bias correction method by Hagmann and Scaillet (2007) can be viewed as an asymmetric kernel version of the semiparametric bias correction method studied by Hjort and Glad (1995) and Hjort and Jones (1996). While Hagmann and Scaillet (2007) focus exclusively on bias correction in the original data scale on $\mathbb{R}_+$, Gustafsson et al. (2009) correct the bias of the initial parametric density estimate on

$\mathbb{R}_+$ after transforming the original data into those on [0, 1] in the spirit of Rudemo (1991). Because of this nature, the approach is called local transformation bias correction. These approaches are designed primarily for misspecification-robust density estimation via nonparametric kernel smoothing, whereas they do not accelerate the bias convergence unlike fully nonparametric bias correction methods. As a result, their best-possible MISE convergence is still $O\left(n^{-4/5}\right)$ in general. Then, following Jones et al. (1999), Hirukawa and Sakudo (2018) combine the semiparametric density estimator by Hjort and Glad (1995) with the bias correction method by Jones et al. (1995) to improve the MISE convergence to $O\left(n^{-8/9}\right)$ in the best-case scenario.

## 3.2 Nonparametric Bias Correction

### 3.2.1 Additive Bias Correction

To generate fourth-order kernels from a given nonnegative symmetric kernel, Jones and Foster (1993) conduct a comprehensive study of the generalized jackknife methods. The additive bias correction ("ABC") by Igarashi and Kakizawa (2015) can be recognized as an extension of the methods to asymmetric kernel density estimation with support on $\mathbb{R}_+$.

Igarashi and Kakizawa (2015) consider the ABC density estimator using the G kernel. Let $\hat{f}_{j,b}(x)$ and $\hat{f}_{j,b/c}(x)$ denote density estimators using the kernel $j$ and smoothing parameters $b$ and $b/c$, respectively, where $c \in (0, 1)$ is some predetermined constant that does not depend on the design point $x$. The ABC estimator using the G kernel is defined as

$$\tilde{f}_{ABC,G}(x) := \frac{1}{1-c}\hat{f}_{G,b}(x) - \frac{c}{1-c}\hat{f}_{G,b/c}(x).$$

In addition, Igarashi and Kakizawa (2015) derive the limit case with $c \uparrow 1$ of this estimator as

$$\check{f}_{ABC,G}(x) : = \lim_{c \uparrow 1}\tilde{f}_{ABC,G}(x)$$

$$= \hat{f}_{G,b}(x) - b\frac{\partial}{\partial b}\hat{f}_{G,b}(x)$$

$$= \frac{1}{n}\sum_{i=1}^{n} K_{G(x,b)}(X_i) H_{G(x,b)}(X_i),$$

where

$$H_{G(x,b)}(u) := 1 - b\frac{\partial}{\partial b}\log K_{G(x,b)}(u)$$

$$= 2 - \left(\frac{u-x}{b}\right) + \frac{x}{b}\left\{\log\left(\frac{u}{b}\right) - \Psi\left(\frac{x}{b}+1\right)\right\}.$$

By construction, the ABC estimator is free of boundary bias, regardless of the value of $c$.

To explore convergence properties of fully nonparametric bias-corrected estimators including the ABC estimator, we impose the regularity conditions below.

**Assumption 3.1**  The fourth-order derivative of the pdf $f(\cdot)$ is continuous and bounded in the neighborhood of $x$.

**Assumption 3.2**  The smoothing parameter $b(=b_n > 0)$ satisfies $b + \left(nb^3\right)^{-1} \to 0$ as $n \to \infty$.

The smoothness condition for the true density $f$ in Assumption 3.1, which is stronger than Assumption 2.1, is standard for consistency of density estimators using fourth-order kernels. Assumption 3.2 implies that the shrinkage rate of the smoothing parameter $b$ must be slower than $O\left(n^{-1/3}\right)$. This condition is required to control the orders of magnitude in remainder terms that appear in the bias approximation to each nonparametrically bias-corrected estimator.

The next theorem documents approximations to the bias and variance of the ABC estimator.

**Theorem 3.1**  (Igarashi and Kakizawa 2015, Theorem 2)
*Under Assumptions 3.1 and 3.2, the bias of $\tilde{f}_{ABC,G}(x)$ can be approximated as*

$$Bias\left\{\tilde{f}_{ABC,G}(x)\right\} = -\frac{1}{c}\left\{f^{(2)}(x) + \frac{5}{6}xf^{(3)}(x) + \frac{x^2}{8}f^{(4)}(x)\right\}b^2 + o\left(b^2\right).$$

*The approximation to the variance of $\tilde{f}_{ABC,G}(x)$ is given by*

$$Var\left\{\tilde{f}_{ABC,G}(x)\right\} = \begin{cases} \frac{1}{nb^{1/2}}\frac{\lambda(c)}{2\sqrt{\pi}}\frac{f(x)}{\sqrt{x}} + o\left(n^{-1}b^{-1/2}\right) & \text{for interior } x \\ O\left(n^{-1}b^{-1}\right) & \text{for boundary } x \end{cases},$$

*where*

$$\lambda(c) := \frac{\left(1+c^{5/2}\right)\left(1+c\right)^{1/2} - 2\sqrt{2}c^{3/2}}{\left(1+c\right)^{1/2}\left(1-c\right)^2}$$

*is monotonously increasing in $c \in (0,1)$ with*

$$\lim_{c\downarrow 0}\lambda(c) = 1 \text{ and } \lim_{c\uparrow 1}\lambda(c) = \frac{27}{16}.$$

*In addition, the bias and variance approximations to $\check{f}_{ABC,G}(x)$ can be obtained by letting $c \uparrow 1$ in the corresponding results.*

ABC improves the bias convergence from $O(b)$ to $O(b^2)$, where the leading bias term can be expressed alternatively as $-(1/c)\mathcal{B}_{2,G}(x,f)b^2$ for $\mathcal{B}_{2,G}(x,f)$ given in Table 2.1. The asymptotic variance of the ABC estimator for interior $x$ is still $O(n^{-1}b^{-1/2})$, whereas the variance coefficient is inflated by a factor of $\lambda(c)$ from that of the G density estimator. It is straightforward to see that orders of magnitude in the MSE-optimal smoothing parameters are $b^*_{ABC,G} = O(n^{-2/9})$ and $b^\dagger_{ABC,G} = O(n^{-1/5})$ for interior and boundary $x$, respectively; the rates are indeed within the range required by Assumption 3.2. Moreover, provided that both $\int_0^\infty \{\mathcal{B}_{2,G}(x,f)\}^2\,dx$ and $\int_0^\infty \{f(x)/\sqrt{x}\}\,dx$ are finite, the MISE of the ABC estimator becomes $O(b^4 + n^{-1}b^{-1/2})$, which yields the MISE-optimal smoothing parameter $b^{**}_{ABC,G} = O(n^{-2/9})$. It follows that the optimal MISE is $O(n^{-8/9})$, i.e., ABC can attain rate improvement. Despite these attractive properties, a concern on ABC is that the resulting estimate may be negative. The bias correction techniques in the next section are free of this issue.

## *3.2.2  Multiplicative Bias Correction*

This section focuses on two class of multiplicative bias correction ("MBC") methods using asymmetric kernels. Unlike ABC, each MBC estimator is free of boundary bias and always generates nonnegative density estimates everywhere by construction.

### 3.2.2.1  MBC by Terrell and Scott (1980)

The first class of MBC considered in this section is to construct a multiplicative combination of two density estimators using different smoothing parameters. Terrell and Scott (1980) (abbreviated as "TS" hereinafter) originally propose this idea in the form of a linear combination of the logarithms of two density estimates with some predetermined weights. Later Koshkin (1988) generalizes the idea, and Jones and Foster (1993) incorporate it into their generalized jackknife methods and reinterpret it as an MBC technique. Transplanting the TS-MBC method into asymmetric kernel density estimation dates back to Hirukawa (2010). Subsequently, Hirukawa and Sakudo (2014, 2015), Funke and Kawka (2015), Igarashi and Kakizawa (2015), and Funke and Hirukawa (2017) extend this method in different directions within the framework of asymmetric kernel smoothing.

Using the same $\hat{f}_{j,b}(x)$ and $\hat{f}_{j,b/c}(x)$ for some $c \in (0, 1)$ as in the previous section, the TS-MBC estimator using the kernel $j$ is defined as

$$\tilde{f}_{TS,j}(x) := \left\{\hat{f}_{j,b}(x)\right\}^{1/(1-c)} \left\{\hat{f}_{j,b/c}(x)\right\}^{-c/(1-c)}.$$

The theorem below presents the bias and variance approximations of the TS-MBC estimator.

**Theorem 3.2** (Hirukawa 2010, Theorem 1; Hirukawa and Sakudo 2014, Theorem 1; Hirukawa and Sakudo 2015, Theorem 3)

*If Assumptions 3.1 and 3.2 and $f(x) > 0$ hold, then the bias of $\tilde{f}_{TS,j}(x)$ for $j \in \{B, MB, G, MG, NM\}$ can be approximated as*

$$Bias\left\{\tilde{f}_{TS,j}(x)\right\} = \frac{1}{c}\left[\frac{1}{2}\frac{\left\{\mathcal{B}_{1,j}(x,f)\right\}^2}{f(x)} + \mathcal{B}_{2,j}(x,f)\right]b^2 + o\left(b^2\right)$$

*for $\mathcal{B}_{1,j}(x,f)$ and $\mathcal{B}_{2,j}(x,f)$ given in Table 2.1. The approximation to the variance of $\tilde{f}_{TS,j}(x)$ is given by*

$$Var\left\{\tilde{f}_{TS,j}(x)\right\} = \begin{cases} \frac{1}{nb^{1/2}}\lambda(c)\,v_j g_j(x)f(x) + o\left(n^{-1}b^{-1/2}\right) & \text{for interior } x \\ O\left(n^{-1}b^{-1}\right) & \text{for boundary } x \end{cases}$$

*for $\lambda(c)$ defined in Theorem 3.1.*

As seen in ABC, TS-MBC improves the bias convergence from $O(b)$ to $O\left(b^2\right)$. The leading bias coefficient of the TS-MBC estimator using the MG kernel, for instance, reduces to

$$\frac{1}{c}\left[\frac{1}{2}\frac{\left\{\mathcal{B}_{1,MG}(x,f)\right\}^2}{f(x)} + \mathcal{B}_{2,MG}(x,f)\right]$$
$$= \frac{1}{c}\left[\frac{x^2}{8}\frac{\left\{f^{(2)}(x)\right\}^2}{f(x)} + \frac{x}{3}f^{(3)}(x) + \frac{x^2}{8}f^{(4)}(x)\right]$$

for $x \geq 2b$. Like ABC, the leading variance coefficient of the TS-MBC estimator is also inflated by a factor of $\lambda(c)$ than that of the density estimator using the same kernel. Interestingly, the dominant term in $Var\left\{\tilde{f}_{TS,G}(x)\right\}$ for interior $x$ is the same as that of the ABC estimator. The optimal MISE of $\tilde{f}_{TS,j}(x)$ is again $O\left(n^{-8/9}\right)$, provided that both $\int \left\{\mathcal{B}_{1,j}(x,f)/f(x) + \mathcal{B}_{2,j}(x,f)\right\}^2 dx$ and $\int g_j(x)f(x)\,dx$ are finite, where $\mathcal{B}_{1,j}$ and $\mathcal{B}_{2,j}$ for $j = MB$ and for $j \in \{MG, NM\}$ refer to those for $x \in [2b, 1-2b]$ and for $x \geq 2b$, respectively.

### 3.2.2.2  MBC by Jones et al. (1995)

The second class of MBC in the spirit of Jones et al. (1995) (abbreviated as "JLN" hereinafter) is based on the identity with respect to the true density $f$

$$f(x) \equiv g(x) \left\{ \frac{f(x)}{g(x)} \right\} := g(x) r(x), \tag{3.1}$$

where $g(\cdot)$ and $r(\cdot)$ are an initial density estimator and a correction factor, respectively. This identity and the kernel $j$ lead to a density estimator of the form

$$\tilde{f}_j(x) = g(x) \left\{ \frac{1}{n} \sum_{i=1}^{n} \frac{K_{j(x,b)}(X_i)}{g(X_i)} \right\}. \tag{3.2}$$

Observe that $g(\cdot) \equiv 1$ (i.e., (improper) uniform density) gives the usual density estimator $\hat{f}(x)$. The original JLN estimator takes $g(\cdot) = \hat{f}_S(\cdot)$ for some nonnegative symmetric kernel and uses the same kernel in place of $K_{j(x,b)}(\cdot)$. This MBC technique is also applied in nonparametric regression estimation (Linton and Nielsen 1994), hazard estimation (Nielsen 1998; Nielsen and Tanggaard 2001), and spectral matrix estimation (Xiao and Linton 2002; Hirukawa 2006). Often applications of JLN-MBC to asymmetric kernel density estimation are investigated on a parallel with TS-MBC; see Hirukawa (2010), Hirukawa and Sakudo (2014, 2015), and Funke and Kawka (2015), for instance.

Substituting $g(\cdot) = \hat{f}_j(\cdot) \left(= \hat{f}_{j,b}(\cdot)\right)$ into (3.2), we can immediately define the JLN-MBC estimator using the kernel $j$ as

$$\tilde{f}_{JLN,j}(x) := \hat{f}_j(x) \left\{ \frac{1}{n} \sum_{i=1}^{n} \frac{K_{j(x,b)}(X_i)}{\hat{f}_j(X_i)} \right\}.$$

The bias and variance approximations of the JLN-MBC estimator are documented in the next theorem.

**Theorem 3.3** (Hirukawa 2010, Theorem 2; Hirukawa and Sakudo 2014, Theorem 2; Hirukawa and Sakudo 2015, Theorem 3)

*If Assumptions 3.1 and 3.2 and $f(x) > 0$ hold, then the bias of $\tilde{f}_{JLN,j}(x)$ for $j \in \{B, MB, G, MG, NM\}$ can be approximated as*

$$Bias \left\{ \tilde{f}_{JLN,j}(x) \right\} = -f(x) \mathcal{B}_{1,j} \left\{ x, h_j(x,f) \right\} b^2 + o\left(b^2\right),$$

*where $\mathcal{B}_{1,j} \left\{ x, h_j(x,f) \right\}$ can be obtained by replacing $f = f(x)$ in $\mathcal{B}_{1,j}(x,f)$ given in Table 2.1 with $h_j(x,f) := \mathcal{B}_{1,j}(x,f)/f(x)$. In addition, $Var \left\{ \tilde{f}_{JLN,j}(x) \right\} \sim Var \left\{ \hat{f}_j(x) \right\}$ regardless of the position of $x$.*

JLN-MBC is yet another method that can improve the bias convergence from $O(b)$ to $O\left(b^2\right)$. The expression $\mathcal{B}_{1,j}\left(x, h_j\right)$ in the leading bias coefficient looks complex at first glance. It should read as follows: for example, if we consider the case of the MG kernel again, then

$$\mathcal{B}_{1,MG}\{x, h_{MG}(x,f)\} = \mathcal{B}_{1,MG}\left\{x, \frac{\mathcal{B}_{1,MG}(x,f)}{f(x)}\right\} = \frac{x}{2}\left\{\frac{x f^{(2)}(x)}{2 \; f(x)}\right\}^{(2)}$$

for $x \geq 2b$, and so on. Also observe that the dominant term in $Var\left\{\tilde{f}_{JLN,j}(x)\right\}$ for interior $x$ is first-order asymptotically equivalent to that of the usual density estimator using the same kernel. This is a sharp contrast to the variances of the ABC and TS-MBC estimators. The optimal MISE of $\tilde{f}_{JLN,j}(x)$ is again $O\left(n^{-8/9}\right)$, provided that both $\int f^2(x)\,\mathcal{B}^2_{1,j}\{x, h_j(x,f)\}\,dx$ and $\int g_j(x) f(x)dx$ are finite, where the choice of $\mathcal{B}_{1,j}$ for $j \in \{MB, MG, NM\}$ obeys the one given below Theorem 3.2.

*Remark 3.1* It is interesting to compare the dominant terms in the bias and variance of $\tilde{f}_{JLN,j}(x)$ with those using a nonnegative symmetric kernel. The original JLN estimator using a nonnegative symmetric kernel is given by

$$\tilde{f}_{JLN,S}(x) = \hat{f}_S(x)\left[\frac{1}{nh}\sum_{i=1}^{n}\frac{K\{(X_i - x)/h\}}{\hat{f}_S(X_i)}\right].$$

It follows from Theorem 1 of Jones et al. (1995) that its bias and variance are

$$Bias\left\{\tilde{f}_{JLN,S}(x)\right\} = -\frac{\mu_2^2}{4}f(x)\left\{\frac{f^{(2)}(x)}{f(x)}\right\}^{(2)}h^4 + o\left(h^4\right), \text{ and}$$

$$Var\left\{\tilde{f}_{JLN,S}(x)\right\} = \frac{1}{nh}R(T_K)f(x) + o\left(n^{-1}h^{-1}\right),$$

where $R(T_K) = \int_{-\infty}^{\infty}T_K^2(u)\,du$ is the roughness of the fourth-order, "twiced" kernel $T_K(u) := 2K(u) - K * K(u)$ by Stuetzle and Mittal (1979). Observe that $\mathcal{B}_{1,j}(x, h_j)$ in $Bias\left\{\tilde{f}_{JLN,j}(x)\right\}$ tends to be much more complex than the corresponding $f(x)\{f^{(2)}(x)/f(x)\}^{(2)}$ in $Bias\left\{\tilde{f}_{JLN,S}(x)\right\}$. Because moments of asymmetric kernels around the design point $x$ are often $O(b)$ or $O(b^2)$, $\mathcal{B}_{1,j}(x, h_j)$ is likely to include extra density derivatives. In addition, while $Var\left\{\tilde{f}_{JLN,j}(x)\right\}$ is first-order asymptotically equivalent to $Var\left\{\hat{f}_j(x)\right\}$, the dominant term in $Var\left\{\tilde{f}_{JLN,S}(x)\right\}$ tends to be larger than the one in $Var\left\{\hat{f}_S(x)\right\}$ due to the fact that $R(T_K) > R(K)$.

*Remark 3.2* Some readers may wonder why TS- and JLN-MBC methods are applied not to the entire family of the GG kernels but individually to two special cases, namely the MG and NM kernels. There are two main reasons. First, whether the bias convergence may speed up from $O(b)$ to $O(b^2)$ depends crucially on whether the second-order term in $Bias\left\{\hat{f}_{GG}(x)\right\}$ is $O(b^2)$. It is worth noting that Conditions 1–5 provide no guidance on the order of magnitude in the second-order bias term. For instance, the second-order term in $Bias\left\{\hat{f}_W(x)\right\}$ is found to be $O\left(b^{3/2}\right)$. Because these MBC techniques merely improve the bias convergence up to $O\left(b^{3/2}\right)$, Theorems 3.2 and 3.3 exclude such inferior cases. The same reason applies to Table

2.1, where the cells of "$\mathcal{B}_{2,j}(x,f)$" for the W kernel are intentionally left blank. Second, to provide the variance approximation of each MBC estimator, we must additionally specify the functional form of $(\alpha, \beta, \gamma) = (\alpha_b(x), \beta_b(x), \gamma_b(x))$. Again Conditions 1–5 alone do not suffice for this purpose.

### 3.2.2.3 Normalization

Neither $\tilde{f}_{TS,j}(x)$ nor $\tilde{f}_{JLN,j}(x)$ integrates to one. In general, MBC leads to lack of normalization, even when nonnegative symmetric kernels are employed. Indeed, Jones et al. (1995, Sect. 2.2) are concerned on this issue in their original JLN estimator and recommend renormalization. The renormalized TS- and JLN-MBC estimators are defined as

$$\tilde{f}_{TS,j}^R(x) := \frac{\tilde{f}_{TS,j}(x)}{\int \tilde{f}_{TS,j}(x)\,dx} \text{ and } \tilde{f}_{JLN,j}^R(x) := \frac{\tilde{f}_{JLN,j}(x)}{\int \tilde{f}_{JLN,j}(x)\,dx},$$

respectively. Notice that the renormalization procedure is equivalent to the "macro" approach by Gouriéroux and Monfort (2006) described in Chap. 2. It is straightforward to see that the biases of these renormalized estimators are

$$Bias\left\{\tilde{f}_{TS,j}^R(x)\right\}$$
$$= \frac{1}{c}\left[\left\{\frac{\mathcal{B}_{1,j}(x,f)}{f(x)} + \mathcal{B}_{2,j}(x,f)\right\} - \int\left\{\frac{\mathcal{B}_{1,j}(x,f)}{f(x)} + \mathcal{B}_{2,j}(x,f)\right\}dx\right]b^2$$
$$+ o\left(b^2\right), \text{ and }$$
$$Bias\left\{\tilde{f}_{JLN,j}^R(x)\right\}$$
$$= -\left\{f(x)\,\mathcal{B}_{1,j}\left(x,h_j\right) - \int f(x)\,\mathcal{B}_{1,j}\left(x,h_j\right)dx\right\}b^2 + o\left(b^2\right),$$

provided that each integrand is absolutely integrable, where the choices of $\mathcal{B}_{1,j}$ and $\mathcal{B}_{2,j}$ for $j \in \{MB, MG, NM\}$ in integrands again obey those given below Theorem 3.2. On the other hand, their asymptotic variances are unaffected.

### 3.2.2.4 Further Bias Reduction via Iteration

In principle, further bias reduction is possible after the regularity conditions are properly strengthened. Constructing a multiplicative combination of $(s+1)$ different density estimators, we can generalize the TS-MBC estimator $\tilde{f}_{TS,j}(x)$ as

$$\tilde{f}_{TS,j}^{(s)}(x) = \prod_{r=0}^{s}\left\{\hat{f}_{j,b/c_r}(x)\right\}^{\alpha_{s,r}},$$

where $c_0 = 1, c_1, \ldots, c_s \in (0, 1)$ are mutually different constants, and the exponent is

$$\alpha_{s,r} = \frac{(-1)^s c_r^s}{\displaystyle\prod_{p=0, p \neq r}^{s} (c_p - c_r)}.$$

Similarly, the $s$th iterated JLN-MBC estimator can be defined as

$$\tilde{f}_{JLN,j}^{(s)}(x) = \tilde{f}_{JLN,j}^{(s-1)}(x) \left\{ \frac{1}{n} \sum_{i=1}^{n} \frac{K_{j(x,b)}(X_i)}{\tilde{f}_{JLN,j}^{(s-1)}(X_i)} \right\},$$

where $\tilde{f}_{JLN,j}^{(0)}(x) = \hat{f}_j(x)$. If $f(x) > 0$, the $2(s+1)$th-order derivative of $f(\cdot)$ is continuous and bounded in the neighborhood of $x$, and the smoothing parameter $b$ satisfies $b + 1/(nb^{2s+1}) \to 0$, then it can be demonstrated that the bias of each estimator is accelerated to $O(b^{s+1})$ while its variance remains $O(n^{-1}b^{-1/2})$ and $O(n^{-1}b^{-1})$ for interior and boundary $x$, respectively. In particular, $Var\left\{\tilde{f}_{JLN,j}^{(s)}(x)\right\}$ is shown to be first-order asymptotically equivalent to $Var\left\{\tilde{f}_{JLN,j}^{(0)}(x)\right\} = Var\left\{\hat{f}_j(x)\right\}$ for interior $x$. Their optimal MSEs are $O\left\{n^{-(4s+4)/(4s+5)}\right\}$ and $O\left\{n^{-(2s+2)/(2s+3)}\right\}$ for interior and boundary $x$, respectively. Accordingly, as the number of iterations increases, global convergence rates of the iterated MBC estimators when best implemented can be arbitrarily close to the parametric one. However, it is doubtful whether there is much gain in practice from these estimators, and thus the iterative procedure is not pursued any further.

### 3.2.2.5 Bias Correction in Joint Density Estimation

Often it is of interest to estimate a joint pdf using multivariate bounded data. While Jones et al. (1995, Sect. 4.1) have already considered a multivariate extension of their original JLN estimation, it is very recently that Funke and Kawka (2015) extend the two MBC techniques to joint density estimation using asymmetric kernels. Suppose that we estimate the joint density $f$ with support on $[0, 1]^d$ or $\mathbb{R}_+^d$ using a random sample $\{\mathbf{X}_i\}_{i=1}^n = \left\{(X_{1i}, \ldots, X_{di})^\top\right\}_{i=1}^n$. Following Bouezmarni and Rombouts (2010), Funke and Kawka (2015) adopt the product kernel method, where

$$\mathbb{K}_{j(\mathbf{x},b)}(\mathbf{u}) := \prod_{\ell=1}^{d} K_{j(x_\ell,b)}(u_\ell)$$

is the product asymmetric kernel given a design point $\mathbf{x} = (x_1, \ldots, x_d)^\top$ and a common smoothing parameter $b_1 = \cdots = b_d = b$. Then, the joint density estimator using the product kernel $\mathbb{K}_{j(\mathbf{x},b)}(\cdot)$ is given by

$$\hat{f}_j(\mathbf{x}) = \hat{f}_{j,b}(\mathbf{x}) = \frac{1}{n} \sum_{i=1}^{n} \mathbb{K}_{j(\mathbf{x},b)}(\mathbf{X}_i).$$

The corresponding multivariate TS- and JLN-MBC estimators can be defined as

$$\tilde{f}_{TS,j}(\mathbf{x}) := \left\{\hat{f}_{j,b}(\mathbf{x})\right\}^{1/(1-c)} \left\{\hat{f}_{j,b/c}(\mathbf{x})\right\}^{-c/(1-c)}$$

for some $c \in (0, 1)$, and

$$\tilde{f}_{JLN,j}(\mathbf{x}) := \hat{f}_j(\mathbf{x}) \left\{\frac{1}{n} \sum_{i=1}^{n} \frac{\mathbb{K}_{j(\mathbf{x},b)}(\mathbf{X}_i)}{\hat{f}_j(\mathbf{X}_i)}\right\}.$$

If $f(\mathbf{x}) > 0$, all fourth-order partial derivatives of $f(\cdot)$ are continuous and bounded in the neighborhood of $\mathbf{x}$, and the smoothing parameter $b$ satisfies $b + \left(nb^{d+2}\right)^{-1} \to 0$, then each estimator is shown to have a $O\left(b^2\right)$ bias and a $O\left\{n^{-1} \prod_{\ell=1}^{d} b^{-(1/2)(1+\mathbf{1}_\ell)}\right\}$ variance, where $\mathbf{1}_\ell := \mathbf{1}\{x_\ell/b \to \kappa_\ell \in (0, \infty)\}$. For interested readers, analytical expressions of the leading bias and variance terms can be found in Theorems 2.1 and 2.2 of Funke and Kawka (2015).

## 3.3 Semiparametric Bias Correction

### 3.3.1 Local Multiplicative Bias Correction

To estimate the unknown density $f$ with support on $\mathbb{R}_+$, Hagmann and Scaillet (2007) consider a local model

$$m\{x, \boldsymbol{\theta}_1, \boldsymbol{\theta}_2(x)\} := f(x, \boldsymbol{\theta}_1) r\{x, \boldsymbol{\theta}_2(x)\}. \tag{3.3}$$

Notice that $f(\cdot, \boldsymbol{\theta}_1)$ is an initial parametric density with the global parameter $\boldsymbol{\theta}_1 \in \boldsymbol{\Theta}_1 \subseteq \mathbb{R}^p$ that is supposed to be close to $f$ but subject to misspecification. In addition, $r\{\cdot, \boldsymbol{\theta}_2(\cdot)\}$ serves as a local parametric model with the local parameter $\boldsymbol{\theta}_2(\cdot) \in \boldsymbol{\Theta}_2 \subseteq \mathbb{R}^q$ that is also an estimate of the correction factor $r(\cdot) = f(\cdot)/f(\cdot, \boldsymbol{\theta}_1)$. Observe that this setup corresponds to the identity (3.1) with $g(\cdot) = f(\cdot, \boldsymbol{\theta}_1)$ plugged in and $r(\cdot)$ replaced by $r\{\cdot, \boldsymbol{\theta}_2(\cdot)\}$.

The estimation procedure of the local model $m$ takes two steps. The first step estimates the initial parametric model by ML in the spirit of Hjort and Glad (1995). Let $\hat{\boldsymbol{\theta}}_1$ be the maximum likelihood estimate ("MLE") of $\boldsymbol{\theta}_1$. The second step is designed to reduce the bias caused by possible misspecification of the parametric start $f(\cdot, \boldsymbol{\theta}_1)$. For this purpose, $\boldsymbol{\theta}_2(x)$ can be estimated via the local likelihood method in the spirit of Hjort and Jones (1996). Specifically, $\hat{\theta}_{2,G}(x)$, the local likelihood estimator using the G kernel, is defined as the solution to the system of $q$ equations $\mathbf{V}_n\left(x, \hat{\theta}_1, \theta_2\right) = \mathbf{0}$, where

$$\mathbf{V}_n\left(x, \hat{\theta}_1, \theta_2\right) := \frac{1}{n}\sum_{i=1}^{n} K_{G(x,b)}\left(X_i\right)\mathbf{v}\left(x, X_i, \theta_2\right)$$

$$- \int_0^\infty K_{G(x,b)}\left(t\right)\mathbf{v}\left(x, t, \theta_2\right)m\left(t, \hat{\theta}_1, \theta_2\right)dt \qquad (3.4)$$

for the general weight function $\mathbf{v}\left(x, t, \theta_2\right) \in \mathbb{R}^q$. If the score $\partial\log r\left(t, \theta_2\right)/\partial\theta_2$ is chosen as the weight function, then $\mathbf{V}_n\left(x, \hat{\theta}_1, \theta_2\right) = \mathbf{0}$ reduces to the first-order condition of the original local log-likelihood by Hjort and Jones (1996). The bias correction method by Hagmann and Scaillet (2007) is called local multiplicative bias correction ("LMBC") in the sense that the correction factor $r$ is modeled and estimated locally. The LMBC estimator using the G kernel is finally defined as

$$\tilde{f}_{LMBC,G}(x) := f\left(x, \hat{\theta}_1\right)r\left\{x, \hat{\theta}_{2,G}(x)\right\}.$$

To describe convergence properties of the LMBC estimator, we introduce additional notations. Let $\theta_1^0$ be the pseudo-true value which minimizes the Kullback–Leibler distance of $f\left(x, \theta_1\right)$ from the true $f(x)$. We also denote $f_0\left(\cdot\right) := f\left(\cdot, \theta_1^0\right)$ and $r_0\left(\cdot\right) := f\left(\cdot\right)/f\left(\cdot, \theta_1^0\right)$. Moreover, define

$$\mathbf{V}\left(x, \theta_1^0, \theta_2\right) := \int_0^\infty K_{G(x,b)}\left(t\right)\mathbf{v}\left(x, t, \theta_2\right)f_0\left(t\right)\left\{r_0\left(t\right) - r\left(t, \theta_2\right)\right\}dt,$$

and let $\theta_2^0$ be the solution to $\mathbf{V}\left(x, \theta_1^0, \theta_2\right) = \mathbf{0}$. After stating an additional regularity condition on the initial parametric model, we present a theorem on the bias and variance approximations to the LMBC estimator.

**Assumption 3.3 (i)** The parameter space $\Theta_1$ is a compact subset of $\mathbb{R}^p$. **(ii)** $f\left(x; \theta_1\right)$ is continuos in $\theta_1$ for every $x \in \mathbb{R}_+$. **(iii)** $\theta_1^0$ is interior in $\Theta_1$ and uniquely minimizes the Kullback-Leibler distance. **(iv)** The log-likelihood of the sample $\{X_i\}_{i=1}^n$ uniformly converges in probability to the Kullback-Leibler distance. **(v)** $\sqrt{n}\left(\hat{\theta}_1 - \theta_1^0\right) = O_p\left(1\right)$.

**Theorem 3.4** (Hagmann and Scaillet 2007, Proposition 1)

*If Assumption 2.2 and 3.3 hold, $\mathbf{V}_n\left(x, \hat{\theta}_1, \theta_2\right) \overset{p}{\to} \mathbf{V}\left(x, \theta_1^0, \theta_2\right)$ as $n \to \infty$, $\theta_2^0$ uniquely solves $\mathbf{V}\left(x, \theta_1^0, \theta_2\right) = \mathbf{0}$, and the second-order derivatives of $r_0\left(\cdot\right)$ and $r\left(\cdot, \theta_2^0\right)$ are continuous and bounded in the neighborhood of $x$, then for $q \geq 2$,*

$$Bias\left\{\tilde{f}_{LMBC,G}(x)\right\} = \frac{x}{2}f_0(x)\left\{r_0^{(2)}(x) - r^{(2)}\left(x, \theta_2^0\right)\right\}b + o\left(b\right).$$

*In addition, $Var\left\{\tilde{f}_{LMBC,G}(x)\right\} \sim Var\left\{\hat{f}_G(x)\right\}$ regardless of the position of x.*

As explained in Hjort and Jones (1996), Hagmann and Scaillet (2007), the leading bias coefficient would include extra terms with first-order derivatives of $f_0(x)$, $r_0(x)$ and $r\left(x, \theta_2^0\right)$ if the dimension of the local parameter $\theta_2$ were limited to one. Therefore, it is recommended that two or more local parameters are fitted in the second step. Inspecting the leading term of *Bias* $\left\{\tilde{f}_{LMBC,G}(x)\right\}$ in this case also reveals that the term becomes small if the local model of the correction factor $r\left(\cdot, \theta_2^0\right)$ can locally capture the curvature of the true correction factor $r_0(\cdot)$ so that $r_0^{(2)}(\cdot) \approx r^{(2)}\left(\cdot, \theta_2^0\right)$ in the neighborhood of $x$. Furthermore, if the model is correct, then the local likelihood estimator is unbiased up to the order considered.

To implement the LMBC estimation, Hagmann and Scaillet (2007) suggest choosing a gamma pdf and the local log-linear regression model in the first and second steps, respectively; see Example 2 of Hagmann and Scaillet (2007) for more details. It can be also found that like the JLN-MBC estimation using the G kernel, the variance of the LMBC estimator is first-order asymptotically equivalent to that of $\hat{f}_G(x)$.

### 3.3.2 Local Transformation Bias Correction

Gustafsson et al. (2009) propose yet another semiparametric MBC technique called local transformation bias correction ("LTBC"), which basically follows the idea of Rudemo (1991). LMBC and LTBC specialize in density estimation with support on $\mathbb{R}_+$, and in each approach an asymmetric kernel plays an important role in reducing the bias of the initial parametric density estimate nonparametrically. A key difference between two approaches is that LMBC and LTBC correct the bias in the original and transformed data scales, respectively.

Although LTBC also relies on the local model (3.3), the bias correction is made in the scale on [0, 1] implied by the probability integral transform. A rationale for LTBC is that if the cumulative distribution function ("cdf") of the initial parametric distribution were the true cdf $F$, then the correction term $r$ would be the pdf of $U[0, 1]$. To rewrite the local model (3.3) suitably, let

$$F_{\theta_1}(x) := F(x, \theta_1) = \int_0^x f(t, \theta_1)\, dt$$

be the cdf of the initial parametric distribution indexed by the global parameter $\theta_1$. The local model for LTBC can be expressed eventually as

$$m\left\{F_{\theta_1}(x), \theta_1, \theta_2\left(F_{\theta_1}(x)\right)\right\} = F_{\theta_1}^{(1)}(x)\, r\left\{F_{\theta_1}(x), \theta_2\left(F_{\theta_1}(x)\right)\right\}$$
$$= f(x, \theta_1)\, r\left\{F_{\theta_1}(x), \theta_2\left(F_{\theta_1}(x)\right)\right\}.$$

The estimation procedure of the local model $m$ again takes two steps. In the first step, $\theta_1$ can be estimated by ML. For the MLE $\hat{\theta}_1$, denote the transformed data and design points by $\left(\hat{U}, \hat{u}\right) := \left(F_{\hat{\theta}_1}(X), F_{\hat{\theta}_1}(x)\right)$. Because their supports are on [0, 1], the second, bias correction step applies the B kernel. Then, $\hat{\theta}_{2,B}(\hat{u})$, the local likelihood estimator using the B kernel, solves the system of $q$ equations $\mathbf{V}_n\left\{\hat{u}, \hat{\theta}_1, \theta_2(\hat{u})\right\} = \mathbf{0}$, where

$$
\mathbf{V}_n\left\{\hat{u}, \hat{\theta}_1, \theta_2(\hat{u})\right\} := \frac{1}{n}\sum_{i=1}^{n} K_{B(\hat{u},b)}\left(\hat{U}_i\right) \mathbf{v}\left(\hat{u}, \hat{U}_i, \theta_2\right)
$$

$$
- \int_0^1 K_{B(\hat{u},b)}(t)\, \mathbf{v}\left(\hat{u}, t, \theta_2\right) r(t, \theta_2)\, dt
$$

for the general weight function $\mathbf{v}(\hat{u}, t, \theta_2) \in \mathbb{R}^q$. As before, if the score $\partial \log r(t, \theta_2)/\partial \theta_2$ is chosen as the weight function, then $\mathbf{V}_n\left\{\hat{u}, \hat{\theta}_1, \theta_2(\hat{u})\right\} = \mathbf{0}$ is the first-order condition of the original local log-likelihood by Hjort and Jones (1996). The LTBC estimator using the B kernel is finally defined as

$$
\tilde{f}_{LTBC,B}(x) := f\left(x, \hat{\theta}_1\right) r\left\{\hat{u}, \hat{\theta}_{2,B}(\hat{u})\right\}.
$$

To describe convergence properties of the LTBC estimator, let $u^0 := F_{\theta_1^0}(x)$ for the pseudo-true value $\theta_1^0$. Also let $r(\cdot)$ be the true pdf of the transformed data $F_{\theta_1}(X)$. Furthermore, define

$$
\mathbf{V}\left\{u, \theta_1^0, \theta_2(u)\right\} := \int_0^1 K_{B(u,b)}(t)\, \mathbf{v}(x, t, \theta_2)\left\{r(t) - r(t, \theta_2)\right\} dt,
$$

and let $\theta_2^0(u^0)$ be the solution to $\mathbf{V}\left\{u^0, \theta_1^0, \theta_2(u^0)\right\} = \mathbf{0}$. Using $f_0(\cdot) = f(\cdot, \theta_1^0)$, the following theorem documents the bias and variance approximations to the LTBC estimator.

**Theorem 3.5** (Gustafsson et al. 2009, Proposition 1)
*If Assumption 2.2 and 3.3 hold, $\mathbf{V}_n\left\{\hat{u}, \hat{\theta}_1, \theta_2(\hat{u})\right\} \xrightarrow{p} \mathbf{V}\left\{u, \theta_1^0, \theta_2(u)\right\}$ as $n \to \infty$, $\theta_2^0(u^0)$ uniquely solves $\mathbf{V}\left\{u^0, \theta_1^0, \theta_2(u^0)\right\} = \mathbf{0}$, and the second-order derivatives of $r(\cdot)$ and $r(\cdot, \theta_2^0(\cdot))$ are continuous and bounded in the neighborhood of $u^0$, then for $q \geq 2$,*

$$
Bias\left\{\tilde{f}_{LTBC,B}(x)\right\}
$$
$$
= \frac{1}{2}f_0(x)\, u^0\left(1 - u^0\right)\left\{r^{(2)}\left(u^0\right) - r^{(2)}\left(u^0, \theta_2^0(u^0)\right)\right\} b + o(b).
$$

*In addition,*

$$Var\left\{\tilde{f}_{LTBC,B}(x)\right\}$$

$$= \begin{cases} \frac{1}{nb^{1/2}}\left(\frac{1}{2\sqrt{\pi}}\right)\frac{f_0(x)f(x)}{\sqrt{u^0(1-u^0)}} + o\left(n^{-1}b^{-1/2}\right) \text{ for interior } u^0 \\ \frac{1}{nb}\frac{\Gamma(2\kappa+1)}{2^{2\kappa+1}\Gamma^2(\kappa+1)}f_0(x)f(x) + o\left(n^{-1}b^{-1}\right) \text{ for boundary } u^0 \end{cases}.$$

As mentioned in the previous section, a correction factor with two or more local parameters should be considered for a better bias property. Also observe that the dominant term in $Var\left\{\tilde{f}_{LTBC,B}(x)\right\}$ differs from the one in $Var\left\{\hat{f}_B(x)\right\}$. Gustafsson et al. (2009) explain that the extra term $f_0(x)$ appears because the probability integral transform induces an implicit location-dependent smoothing parameter $b/\{f_0(x)\}^2$.

Comparing LTBC with LMBC in terms of computational aspects, Gustafsson et al. (2009) also argue that the local likelihood step for LMBC is independent of the global parametric start, which may be considered as a disadvantage. On the other hand, LTBC has no closed form, and thus it must always rely on numerical approximations. Certain combinations of a global parametric start, a local model of the correction factor and a kernel, can yield simple closed-form expressions of LMBC; see Sect. 4.4 of Hagmann and Scaillet (2007) for more details.

To implement the LTBC estimation, Gustafsson et al. (2009) recommend choosing the pdf of the generalized Champernowne distribution by Buch-Larsen et al. (2005) and the local log-linear density model in the first and second steps, respectively. Because of its Pareto-type tail, the generalized Champernowne distribution is useful for modeling distributions of insurance payments and operational risks.

### 3.3.3 Rate Improvement via Combining with JLN-MBC

The initial parametric model for LMBC and LTBC is misspecified in almost all cases. As a consequence, the bias convergence of these estimators remains at $O(b)$ in general. Within the framework of symmetric kernel smoothing, Jones et al. (1999) (abbreviated as "JSH" hereinafter) combine the semiparametric density estimator by Hjort and Glad (1995) (abbreviated as "HG" hereinafter) with JLN-MBC. JSH's approach aims at acquiring the best aspects of both parametric and nonparametric density estimation. When the parametric start for HG estimation is close enough to the true density, the JSH-MBC estimator can attain considerable efficiency at the stage of parametric fitting, as in LMBC and LTBC. Even if the parametric start is shown not to be good enough, additional JLN-type bias correction can still generate an estimator with bias of smaller order.

Hirukawa and Sakudo (2018) extend this approach to asymmetric kernel density estimation. Applying (3.2) repeatedly constitutes the entire JSH-MBC estimation. In the first step, consider an initial parametric model $f(\cdot) = f(\cdot, \theta_1)$ and estimate $\theta_1$ by ML, as in LMBC and LTBC. In the second step, substituting $g(\cdot) = f\left(\cdot, \hat{\theta}_1\right)$ for the MLE $\hat{\theta}_1$ into (3.2) yields the HG density estimate using the kernel $j$

$$\tilde{f}_{HG,j}(x) := f\left(x; \hat{\boldsymbol{\theta}}_1\right) \left\{ \frac{1}{n} \sum_{i=1}^{n} \frac{K_{j(x,b)}(X_i)}{f\left(X_i; \hat{\boldsymbol{\theta}}_1\right)} \right\}.$$

Then, in the final step, putting $g\left(\cdot\right) = \tilde{f}_{HG,j}\left(\cdot\right)$ again in (3.2), we can define the JSH-MBC density estimator using the kernel $j$ as

$$\tilde{f}_{JSH,j}(x) := \tilde{f}_{HG,j}(x) \left\{ \frac{1}{n} \sum_{i=1}^{n} \frac{K_{j(x,b)}(X_i)}{\tilde{f}_{HG,j}(X_i)} \right\}.$$

The theorem below documents the bias and variance approximations to the JSH-MBC estimator. Notice that the definitions of $\theta_1^0$ and $r_0\left(\cdot\right)$ are the same as used for Theorem 3.4.

**Theorem 3.6** (Hirukawa and Sakudo 2018, Theorem 1)

*If Assumptions 3.2 and 3.3 hold, the fourth-order derivative of $r_0\left(\cdot\right)$ is continuous and bounded in the neighborhood of $x$, and $f\left(x\right), f_0\left(x\right) > 0$, then the bias of $\tilde{f}_{JSH,j}\left(x\right)$ for $j \in \{B, MB, G, MG, NM\}$ can be approximated as*

$$Bias\left\{\tilde{f}_{JSH,j}\left(x\right)\right\} = -f\left(x\right) \mathcal{B}_{1,j}\left\{x, h_j\left(x, r_0\right)\right\} b^2 + o\left(b^2\right),$$

*where $\mathcal{B}_{1,j}\left\{x, h_j\left(x, r_0\right)\right\}$ can be obtained by replacing $f = f\left(x\right)$ in $\mathcal{B}_{1,j}\left(x, f\right)$ given in Table 2.1 with $h_j\left(x, r_0\right) := \mathcal{B}_{1,j}\left(x, r_0\right)/r_0\left(x\right)$. In addition, $Var\left\{\tilde{f}_{JSH,j}\left(x\right)\right\} \sim Var\left\{\hat{f}_j\left(x\right)\right\}$ regardless of the position of $x$.*

The term $\mathcal{B}_{1,j}\left\{x, h_j\left(x, r_0\right)\right\}$ in the leading bias coefficient can be obtained straightforwardly by replacing $f$ in $\mathcal{B}_{1,j}\left\{x, h_j\left(x, f\right)\right\}$ given in Theorem 3.3 with $r_0$. Hence, the former should be read in a similar manner to the latter. Also invoke that $r_0\left(\cdot\right) = f\left(\cdot\right)/f\left(\cdot, \theta_1^0\right)$. It follows that if the initial parametric start $f\left(\cdot, \theta_1^0\right) \equiv 1$ (i.e., if the (improper) uniform density is chosen as the "start"), then the JSH-MBC density estimator collapses to the JLN-MBC estimator using the same kernel. Furthermore,

$$Var\left\{\tilde{f}_{JSH,j}\left(x\right)\right\} \sim Var\left\{\tilde{f}_{JLN,j}\left(x\right)\right\} \sim Var\left\{\hat{f}_j\left(x\right)\right\} \sim \frac{1}{nb^{1/2}} v_j g_j\left(x\right) f\left(x\right)$$

for interior $x$. This contrasts with the original JSH estimator using symmetric kernels in that

$$Var\left\{\tilde{f}_{JSH,s}\left(x\right)\right\} \sim Var\left\{\tilde{f}_{JLN,s}\left(x\right)\right\} \sim \frac{1}{nh} R\left(T_K\right) f\left(x\right)$$

$$> \frac{1}{nh} R\left(K\right) f\left(x\right) \sim Var\left\{\hat{f}_s\left(x\right)\right\}.$$

Lastly (but not least importantly), acceleration of the bias convergence from $O(b)$ to $O(b^2)$ improves the optimal MISE of $\hat{f}_{JSH,j}(x)$ to $O(n^{-8/9})$, provided that both $\int f^2(x) \mathcal{B}^2_{1,j}\{x, h_j(x, r_0)\} dx$ and $\int g_j(x)f(x) dx$ are finite, where, as before, the choice of $\mathcal{B}_{1,j}$ for $j \in \{MB, MG, NM\}$ obeys the one given below Theorem 3.2.

## 3.4 Smoothing Parameter Selection

There are very few works on smoothing parameter selection in bias-corrected estimation. It appears that this problem is hard to resolve even in the case of bias correction in density estimation using nonnegative symmetric kernels. Indeed, Jones and Signorin (1997, Sect. 5) defer automatic bandwidth selection to future work, and there has been not much progress since then, to the best of our knowledge.

As in Chap. 2, methods of choosing the smoothing parameter $b$ can be classified roughly into plug-in and CV approaches. Hirukawa (2010), Hirukawa and Sakudo (2014) develop plug-in methods for TS- and JLN-MBC estimators that are similar to the ones given in Chap. 2. Their analytical expressions are complex in general, and thus we suggest that interested readers look into Sect. 3.1 of Hirukawa (2010) and Sect. 3.2 of Hirukawa and Sakudo (2014).

In contrast, in implementing the LTBC estimator, Gustafsson et al. (2009) employ $\hat{b}_{LTBC} = \hat{\sigma} n^{-2/5}$ for the bias correction step using the B kernel, where $\hat{\sigma}$ is the sample standard deviation of the transformed data on [0, 1]. This particular choice reflects that if the global parametric model is correct, then the transformed data are distributed as $U[0, 1]$. Unbiasedness of $\hat{f}_B(\cdot)$ for the pdf of $U[0, 1]$ (see Remark 2.1) makes it difficult to select the smoothing parameter on the MISE basis. As a result, they decide to employ the simplest possible smoothing parameter. However, the parametric model is typically misspecified, and thus $\hat{b}_{MB}$ in Chap. 2 could also work in this context.

Some authors alternatively consider the CV method. Hagmann and Scaillet (2007) apply it to the LMBC estimation, whereas Funke and Kawka (2015) use it in the context of multivariate TS- and JLN-MBC estimations. More details can be found in Sect. 4.5 of Hagmann and Scaillet (2007) and Sect. 3.3 of Funke and Kawka (2015).

## References

Bouezmarni, T., and J.V.K. Rombouts. 2010. Nonparametric density estimation for multivariate bounded data. *Journal of Statistical Planning and Inference* 140: 139–152.

Buch-Larsen, T., J.P. Nielsen, M. Guillén, and C. Bollancé. 2005. Kernel density estimation for heavy-tailed distributions using the Champernowne transformation. *Statistics* 39: 503–518.

Funke, B., and M. Hirukawa. 2017. Nonparametric estimation and testing on discontinuity of positive supported densities: a kernel truncation approach. *Econometrics and Statistics*, forthcoming.

Funke, B., and R. Kawka. 2015. Nonparametric density estimation for multivariate bounded data using two non-negative multiplicative bias correction methods. *Computational Statistics and Data Analysis* 92: 148–162.

Gouriéroux, C., and A. Monfort. 2006. (Non)Consistency of the beta kernel estimator for recovery rate distribution. CREST Discussion Paper n° 2006–31.

Gustafsson, J., M. Hagmann, J.P. Nielsen, and O. Scaillet. 2009. Local transformation kernel density estimation of loss distributions. *Journal of Business and Economic Statistics* 27: 161–175.

Hagmann, M., and O. Scaillet. 2007. Local multiplicative bias correction for asymmetric kernel density estimators. *Journal of Econometrics* 141: 213–249.

Hirukawa, M. 2006. A modified nonparametric prewhitened covariance estimator. *Journal of Time Series Analysis* 27: 441–476.

Hirukawa, M. 2010. Nonparametric multiplicative bias correction for kernel-type density estimation on the unit interval. *Computational Statistics and Data Analysis* 54: 473–495.

Hirukawa, M., and M. Sakudo. 2014. Nonnegative bias reduction methods for density estimation using asymmetric kernels. *Computational Statistics and Data Analysis* 75: 112–123.

Hirukawa, M., and M. Sakudo. 2015. Family of the generalised gamma kernels: a generator of asymmetric kernels for nonnegative data. *Journal of Nonparametric Statistics* 27: 41–63.

Hirukawa, M., and M. Sakudo. 2018. Another bias correction for asymmetric kernel density estimation with a parametric start. Working Paper.

Hjort, N.L., and I.K. Glad. 1995. Nonparametric density estimation with a parametric start. *Annals of Statistics* 23: 882–904.

Hjort, N.L., and M.C. Jones. 1996. Locally parametric nonparametric density estimation. *Annals of Statistics* 24: 1619–1647.

Igarashi, G., and Y. Kakizawa. 2015. Bias corrections for some asymmetric kernel estimators. *Journal of Statistical Planning and Inference* 159: 37–63.

Jones, M.C., and P.J. Foster. 1993. Generalized jackknifing and higher order kernels. *Journal of Nonparametric Statistics* 3: 81–94.

Jones, M.C., O. Linton, and J.P. Nielsen. 1995. A simple bias reduction method for density estimation. *Biometrika* 82: 327–338.

Jones, M.C., and D.F. Signorini. 1997. A comparison of higher-order bias kernel density estimators. *Journal of the American Statistical Association* 92: 1063–1073.

Jones, M.C., D.F. Signorini, and N.L. Hjort. 1999. On multiplicative bias correction in kernel density estimation. *Sankhyā: The Indian Journal of Statistics, Series A* 61: 422–430.

Koshkin, G.M. 1988. Improved non-negative kernel estimate of a density. *Theory of Probability and Its Applications* 33: 759–764.

Linton, O., and J.P. Nielsen. 1994. A multiplicative bias reduction method for nonparametric regression. *Statistics and Probability Letters* 19: 181–187.

Nielsen, J.P. 1998. Multiplicative bias correction in kernel hazard estimation. *Scandinavian Journal of Statistics* 25: 541–553.

Nielsen, J.P., and C. Tanggaard. 2001. Boundary and bias correction in kernel hazard estimation. *Scandinavian Journal of Statistics* 28: 675–698.

Rudemo, M. 1991. Comment on " Transformations in density estimation" by M. P. Wand, J. S. Marron and D. Ruppert. *Journal of the American Statistical Association* 86: 353–354.

Stuetzle, W., and Y. Mittal. 1979. Some comments on the asymptotic behavior of robust smoothers. In *Smoothing Techniques for Curve Estimation: Proceedings of a Workshop Held in Heidelberg, April 2–4, 1979*, ed. T. Gasser, and M. Rosenblatt, 191–195. Berlin: Springer-Verlag.

Terrell, G.R., and D.W. Scott. 1980. On improving convergence rates for nonnegative kernel density estimators. *Annals of Statistics* 8: 1160–1163.

Xiao, Z., and O. Linton. 2002. A nonparametric prewhitened covariance estimator. *Journal of Time Series Analysis* 23: 215–250.

# Chapter 4
# Regression Estimation

As mentioned in Chap. 1, research in asymmetric kernels appears to begin with regression estimation. This chapter investigates nonparametric regression estimation smoothed by asymmetric kernels. Our primary focus is on the problem of estimating the regression model

$$Y = m(X) + \epsilon, \ E(\epsilon \mid X) = 0,$$

using *i.i.d.* random variables $\{(Y_i, X_i)\}_{i=1}^{n}$ that are assumed to be drawn from a joint distribution with support either on $\mathbb{R} \times [0, 1]$ or $\mathbb{R} \times \mathbb{R}_{+}$. We discuss convergence properties of the Nadaraya–Watson (Nadaraya 1964; Watson 1964) and local linear estimators (Fan and Gijbels 1992; Fan 1993) of the conditional mean function $m$, including their bias and variance approximations and limiting distributions. As an application of regression estimation, the estimation problem of scalar diffusion models is also explained. A brief review of smoothing parameter selection concludes the chapter.

## 4.1 Preliminary

### 4.1.1 The Estimators

The Nadaraya–Watson ("NW") and local linear ("LL") regression smoothers are two popular choices in nonparametric regression estimation. The NW estimator of $m$ at a given design point $x$ using the kernel $j$ can be defined as

$$\hat{m}_j^{nw}(x) := \arg\min_{\beta_0} \sum_{i=1}^{n} (Y_i - \beta_0)^2 K_{j(x,b)}(X_i) = \frac{\sum_{i=1}^{n} Y_i K_{j(x,b)}(X_i)}{\sum_{i=1}^{n} K_{j(x,b)}(X_i)}.$$

© The Author(s) 2018
M. Hirukawa, *Asymmetric Kernel Smoothing*, JSS Research Series
in Statistics, https://doi.org/10.1007/978-981-10-5466-2_4

Often the estimator is alternatively called the local constant estimator, because it is the solution to this optimization problem.

On the other hand, the LL estimator of $m$ at a given design point $x$ using the kernel $j$ is $\hat{m}_j^{ll}(x) = \hat{\beta}_0$, where

$$\left(\hat{\beta}_0, \hat{\beta}_1\right) := \arg\min_{(\beta_0,\beta_1)} \sum_{i=1}^n \{Y_i - \beta_0 - \beta_1 (X_i - x)\}^2 K_{j(x,b)}(X_i).$$

In fact, the LL estimator admits the concise expression

$$\hat{m}_j^{ll}(x) = \frac{S_{2,j}(x) T_{0,j}(x) - S_{1,j}(x) T_{1,j}(x)}{S_{0,j}(x) S_{2,j}(x) - \{S_{1,j}(x)\}^2},$$

where

$$S_{\ell,j}(x) = \sum_{i=1}^n (X_i - x)^\ell K_{j(x,b)}(X_i) \text{ and}$$

$$T_{\ell,j}(x) = \sum_{i=1}^n Y_i (X_i - x)^\ell K_{j(x,b)}(X_i)$$

for $\ell \geq 0$. Observe that $\hat{m}_j^{nw}(x) = T_{0,j}(x) / S_{0,j}(x)$ holds. It is also worth emphasizing that unlike density estimation, lack of normalization in asymmetric kernels is not an issue in regression estimation.

## 4.2   Convergence Properties of the Regression Estimators

### 4.2.1   Regularity Conditions

The regularity conditions below are required to establish asymptotic normality of the regression estimators. Assumptions 4.1 and 4.2 are basically the same as (2.3) of Chen (2002), and they suffice for the bias and variance approximations of the estimators. On the other hand, Assumption 4.3 is required to establish Liapunov's condition for the central limit theorem.

**Assumption 4.1**  The second-order derivative of $m(\cdot)$ is continuous and bounded in the neighborhood of $x$.

**Assumption 4.2**  Let $f(\cdot)$ be the marginal pdf of $X$. Also define $\sigma^2(\cdot) := E(\epsilon|X = \cdot)$. Then, both $f(\cdot)$ and $\sigma^2(\cdot)$ are first-order Lipschitz continuous in the neighborhood of $x$, $f(x) > 0$ and $\sigma^2(x) < \infty$.

**Assumption 4.3** There is some constant $\delta > 0$ so that $E\left(|\epsilon|^{2+\delta}\big|\, X = \cdot\right)$ is bounded uniformly on the support of $X$.

### 4.2.2 Asymptotic Normality of the Estimators

In what follows, our primary focuses are on the NW and LL estimators using the B and G kernels. We consider only a few cases from the following two viewpoints on LL estimation. First, the LL estimators using the B and G kernels eliminate the design bias term including $m^{(1)}(x)\left\{f^{(1)}(x)/f(x)\right\}$, as is the cases with symmetric kernels. This contrasts the fact that the leading biases of $\hat{f}_B(x)$ and $\hat{f}_G(x)$ involve a term with $f^{(1)}(x)$. Second, it is also straightforward to see that for interior $x$, the LL estimators using the MB and MG kernels are first-order asymptotically equivalent to those using the B and G kernels, respectively.

Two theorems below document asymptotic normality of the NW and LL estimators. Each theorem can be demonstrated by making the bias and variance approximations and establishing Liapunov's condition. Asymptotic expansions of kernel regression estimation and moment approximations explained in Chap. 2 directly apply to the bias and variance approximations, whereas Liapunov's condition can be shown in a similar manner to the Proof of Theorem 5.4 in Sect. 5.5.1.

**Theorem 4.1** *Suppose that Assumptions 4.1–4.3 hold. If $b \asymp n^{-2/5}$, then, as $n \to \infty$, for $j \in \{B, G\}$,*

$$\sqrt{nb^{1/2}}\left\{\hat{m}_j^{nw}(x) - m(x) - B_j^{nw}(x)b\right\} \xrightarrow{d} N\left(0, V_j^{nw}(x)\right)$$

*for interior $x$, where*

$$B_B^{nw}(x) = \left\{(1 - 2x) + x(1 - x)\frac{f^{(1)}(x)}{f(x)}\right\}m^{(1)}(x) + \frac{1}{2}x(1 - x)m^{(2)}(x),$$

$$B_G^{nw}(x) = \left\{1 + x\frac{f^{(1)}(x)}{f(x)}\right\}m^{(1)}(x) + \frac{1}{2}xm^{(2)}(x),$$

$$V_B^{nw}(x) = \frac{\sigma^2(x)}{2\sqrt{\pi}\sqrt{x(1 - x)}f(x)},\ and$$

$$V_G^{nw}(x) = \frac{\sigma^2(x)}{2\sqrt{\pi}\sqrt{x}f(x)}.$$

*If $b \asymp n^{-1/3}$, then, as $n \to \infty$, for $j \in \{B, G\}$,*

$$\sqrt{nb}\left\{\hat{m}_j^{nw}(x) - m(x) - m^{(1)}(x)b\right\} \xrightarrow{d} N\left(0, V_{j,0}^{nw}(x)\right)$$

*for $x/b \to \kappa \in (0, \infty)$, and*

$$\sqrt{nb}\left\{\hat{m}_B^{nw}(x) - m(x) + m^{(1)}(x)b\right\} \xrightarrow{d} N\left(0, V_{B,0}^{nw}(x)\right)$$

*for* $(1-x)/b \to \kappa \in (0, \infty)$, *where*

$$V_{j,0}^{nw}(x) = \frac{\Gamma(2\kappa+1)}{2^{2\kappa+1}\Gamma^2(\kappa+1)} \frac{\sigma^2(x)}{f(x)}.$$

**Theorem 4.2** *Suppose that Assumptions 4.1–4.3 hold. If* $b \asymp n^{-2/5}$, *then, as* $n \to \infty$, *for* $j \in \{B, G\}$,

$$\sqrt{nb^{1/2}}\left\{\hat{m}_j^{ll}(x) - m(x) - B_j^{ll}(x)b\right\} \xrightarrow{d} N\left(0, V_j^{nw}(x)\right)$$

*for interior x, and*

$$\sqrt{nb}\left\{\hat{m}_j^{ll}(x) - m(x) - \frac{1}{2}(\kappa-2)m^{(2)}(x)b^2\right\} \xrightarrow{d} N\left(0, \frac{(2\kappa+5)V_{j,0}^{nw}(x)}{2(\kappa+1)}\right)$$

*for* $x/b \to \kappa \in (0, \infty)$ *or* $(1-x)/b \to \kappa$, *where*

$$B_B^{ll}(x) = \frac{1}{2}x(1-x)m^{(2)}(x), \quad B_G^{ll}(x) = \frac{1}{2}xm^{(2)}(x),$$

*and* $V_j^{nw}(x)$ *and* $V_{j,0}^{nw}(x)$ *are defined in Theorem 4.1.*

A few remarkable differences can be found in asymptotic results on the NW and LL estimators. First, as mentioned above, LL estimation eliminates the design bias term. Because this term depends on the distribution of $X$, it is sensitive to the position of the design point $x$. Second, while the bias convergence in NW estimation is $O(b)$ regardless of the position of $x$, the one for boundary $x$ in LL estimation improves to $O(b^2)$. To put it in another way, LL estimation automatically compensates slowdown in the variance convergence for boundary $x$ with acceleration in the bias convergence. As a consequence, the MSE-optimal smoothing parameter and optimal MSE of each LL estimator are $O(n^{-2/5})$ and $O(n^{-4/5})$, respectively, regardless of the position of $x$. In contrast, those of each NW estimator for boundary $x$ remain $O(n^{-1/3})$ and $O(n^{-2/3})$. Observe that Theorems 4.1 and 4.2 rely on the MSE-optimal smoothing parameters. Third, it is easy to see that the optimal MSE of $\hat{m}_B^{ll}(x)$ for interior $x$ and that of $\hat{m}_G^{ll}(x)$ are first-order asymptotically equivalent, i.e.,

$$MSE^*\left\{\hat{m}_B^{ll}(x)\right\} \sim MSE^*\left\{\hat{m}_G^{ll}(x)\right\}$$

$$\sim \frac{5}{4}\left(\frac{1}{4\pi}\right)^{2/5}\left\{m^{(2)}(x)\right\}^{2/5}\left\{\frac{\sigma^2(x)}{f(x)}\right\}^{4/5}n^{-4/5}.$$

The right-hand side is also the optimal MSE of the LL estimator using the Gaussian kernel.

*Remark 4.1* Seifert and Gasser (1996, Theorem 1) argue that finite-sample variances of local polynomial smoothers using kernels with compact support may be unbounded. This problem is typically the case when smoothing is made in sparse regions. As a remedy, Seifert and Gasser (1996) propose to increase the bandwidth locally in such regions. However, Chen (2002, Lemma 1) demonstrates that the $p$th-order local polynomial smoother using the B kernel is immune to this problem, as long as at least $p + 1$ different data points are not on the boundary of $[0, 1]$. It follows that in finite samples $\hat{m}_B^{ll}(\cdot)$ has finite variance with probability 1.

*Remark 4.2* As mentioned in Chap. 1, the B kernel may have been originally proposed in the context of regression estimation. Chen (2000) reports asymptotic properties of Gasser–Müller type regression estimators (Gasser and Müller 1979) using the B and MB kernels. Let $X_1, \ldots, X_n$ be ordered *fixed* regressors so that $0 \le X_1 \le \cdots \le X_n \le 1$. Then, the Gasser–Müller type estimator is defined as

$$\hat{m}_j^{gm}(x) := \sum_{i=1}^{n} Y_i \int_{s_{i-1}}^{s_i} K_{j(x,b)}(u)\,du, \quad j \in \{B, MB\},$$

where $s_i := (X_i + X_{i+1})/2$, $i = 1, \ldots, n - 1$ with $(s_0, s_n) = (0, 1)$. In particular, as reported in Sect. 5 of Chen (2000), $\hat{m}_{MB}^{gm}(x)$ for interior $x$ admits the following bias and variance expansions:

$$Bias\left\{\hat{m}_{MB}^{gm}(x)\right\} = \frac{1}{2}x(1 - x)m^{(2)}(x)b + o(b); \text{ and}$$

$$Var\left\{\hat{m}_{MB}^{gm}(x)\right\} = \frac{1}{nb^{1/2}}\left(\frac{1}{2\sqrt{\pi}}\right)\frac{\sigma^2(x)}{\sqrt{x(1-x)}f(x)} + o\left(n^{-1}b^{-1/2}\right).$$

Observe that the leading bias and variance terms are the same as those of $\hat{m}_B^{ll}(x)$. It follows that the optimal MSE of $\hat{m}_{MB}^{gm}(x)$ for interior $x$ is first-order asymptotically equivalent to those of the LL estimators using the B, G, and Gaussian kernels.

### 4.2.3 Other Convergence Results

As presented in Chap. 2, different types of convergence results are also available for asymmetric kernel regression estimation. Shi and Song (2016, Theorem 3.5) demonstrate uniform strong consistency of the NW estimator using the G kernel with a convergence rate. The uniform convergence may be useful for exploring convergence properties of semiparametric estimation with the estimator employed as a first-stage nonparametric estimator. The result could be also extended to the LL (or even local polynomial) smoother using the G kernel or nonparametric regression estimators using other asymmetric kernels.

### 4.2.4   Regression Estimation Using Weakly Dependent Observations

So far convergence results of nonparametric regression estimators using asymmetric kernels have been developed under *i.i.d.* setting. With growing interest in describing the dynamic behavior of high-frequency financial variables such as the volatility of asset returns and the time duration between market events, more attention has been paid to nonnegative time-series models for past three decades: recent examples include Engle (2002), Gouriéroux and Jasiak (2006) and Brownlees et al. (2012), to name a few. While we may in principle apply asymmetric kernel regression estimation for nonnegative time-series data, we must understand asymptotic properties of the estimators.

Asymptotic results have been already explored for nonparametric regression estimation using symmetric kernels and weakly dependent observations. For example, Masry and Fan (1997) establish asymptotic normality of local polynomial estimators using symmetric kernels for some mixing processes. It appears that asymptotic normality of the regression estimators smoothed by asymmetric kernels for positive $\alpha$-mixing processes can be established by suitably modifying the regularity conditions provided therein. More specifically, it can be demonstrated that Theorems 4.1 and 4.2 still hold under weakly dependent sampling by additionally imposing a battery of conditions, including (i) boundedness of the joint density between two observations, (ii) $\alpha$-mixing of a suitable size, and (iii) a divergence rate of the block size for the small-block and large-block arguments. These conditions correspond to Condition 2 (ii), (iii) and Condition 3 of Masry and Fan (1997), respectively.

## 4.3   Estimation of Scalar Diffusion Models of Short-Term Interest Rates

### 4.3.1   Background

As an interesting application of asymmetric kernel regression estimation, we discuss nonparametric estimation of time-homogeneous drift and diffusion functions in continuous-time models that are used to describe the underlying dynamics of spot interest rates. The general form of the underlying continuous-time process for the spot interest rate $X_t$ is represented by the stochastic differential equation ("SDE")

$$dX_t = \mu(X_t)\,dt + \sigma(X_t)\,dW_t, \tag{4.1}$$

where $W_t$ is a standard Brownian motion, and $\mu(\cdot)$ and $\sigma(\cdot)$ are called the drift and diffusion (or instantaneous volatility) functions, respectively. Stanton (1997) uses the infinitesimal generator and a Taylor series expansion to give the first-order approximations to $\mu(X)$ and $\sigma^2(X)$ as

$$E\left(X_{t+\Delta} - X_t \mid X_t\right) = \mu(X_t)\Delta + o(\Delta) \text{ and}$$
$$E\left\{(X_{t+\Delta} - X_t)^2 \mid X_t\right\} = \sigma^2(X_t)\Delta + o(\Delta),$$

respectively, where $\Delta$ is a discrete, arbitrarily small time step and $o(\Delta)$ denotes the remainder term of a smaller order.

The conditional expectations can be estimated via nonparametric regression smoothing. Although Stanton (1997) originally proposes to use the NW estimation using the Gaussian kernel, Chapman and Pearson (2000) find in their Monte Carlo study that when the true drift is linear in the level of the spot interest rate, there are two biases in Stanton's (1997) drift estimator, namely a bias near the origin and a pronounced downward bias in the region of high interest rates where the data are sparse. This motivates us to incorporate asymmetric kernel smoothing into the Stanton's (1997) nonparametric drift and diffusion estimators.

### 4.3.2 Estimation of Scalar Diffusion Models via Asymmetric Kernel Smoothing

#### 4.3.2.1 The Drift and Diffusion Estimators

Suppose that we observe a discrete sample $\{X_{i\Delta}\}_{i=1}^n$ at $n$ equally spaced time points from the short-rate diffusion process $\{X_t : 0 \leq t \leq T\}$ satisfying the SDE (4.1), where the time span of the sample $T$ and the step size between observations $\Delta$ satisfy $T = n\Delta$. Gospodinov and Hirukawa (2012) focus on nonnegativity of the spot rate $X_t$ and define the NW estimators of drift and diffusion functions using the G kernel for a given design point $x > 0$ as

$$\widehat{\mu}_b(x) := \frac{1}{\Delta} \frac{\sum_{i=1}^{n-1} \left(X_{(i+1)\Delta} - X_{i\Delta}\right) K_{G(x,b)}(X_{i\Delta})}{\sum_{i=1}^{n-1} K_{G(x,b)}(X_{i\Delta})} \text{ and}$$

$$\widehat{\sigma}_b^2(x) := \frac{1}{\Delta} \frac{\sum_{i=1}^{n-1} \left(X_{(i+1)\Delta} - X_{i\Delta}\right)^2 K_{G(x,b)}(X_{i\Delta})}{\sum_{i=1}^{n-1} K_{G(x,b)}(X_{i\Delta})},$$

respectively.

#### 4.3.2.2 Regularity Conditions

Gospodinov and Hirukawa (2012) apply the in-fill and long-span asymptotics such that $n \to \infty$, $T \to \infty$ (long-span), and $\Delta = T/n \to 0$ (in-fill) to deliver convergence properties of $\widehat{\mu}_b(x)$ and $\widehat{\sigma}_b^2(x)$. Exploring the properties requires the following regularity conditions.

**Assumption 4.4** **(i)** $\mu\,(\cdot)$ and $\sigma\,(\cdot)$ are time-homogeneous, $\mathfrak{B}$-measurable functions on $(0,\infty)$, where $\mathfrak{B}$ is the $\sigma$-field generated by Borel sets on $(0,\infty)$. Both functions are at least twice continuously differentiable. Hence, they satisfy local Lipschitz and growth conditions. Thus, for every compact subset $J$ of the range $(0,\infty)$, there exist constants $C_1^J$ and $C_2^J$ such that, for all $x,\,y\in J$,

$$|\mu\,(x) - \mu\,(y)| + |\sigma\,(x) - \sigma\,(y)| \le C_1^J\,|x-y| \text{ and}$$
$$|\mu\,(x)| + |\sigma\,(x)| \le C_2^J\,\{1+|x|\}\,.$$

**(ii)** $\sigma^2\,(\cdot) > 0$ on $(0,\infty)$.
**(iii)** The natural scale function

$$S\,(x) := \int_c^x \exp\left[\int_c^y \left\{-\frac{2\mu\,(u)}{\sigma^2\,(u)}\right\} du\right] dy$$

for some generic constant $c \in (0,\infty)$ satisfies

$$\lim_{x\to 0} S\,(x) = -\infty \text{ and } \lim_{x\to\infty} S\,(x) = \infty.$$

**Assumption 4.5** The speed function

$$s\,(x) := \frac{2}{\sigma^2(x)S^{(1)}\,(x)}$$

satisfies $\int_0^\infty s\,(x)\,dx < \infty$.

**Assumption 4.6** Let the chronological local time of the diffusion process (4.1) be

$$\bar{L}_X\,(T,x) := \frac{1}{\sigma^2\,(x)} \lim_{\epsilon\to 0} \frac{1}{\epsilon} \int_0^T \mathbf{1}\,\{X_s \in [x,x+\epsilon)\}\,\sigma^2\,(X_s)\,ds.$$

Then, as $n,\,T\to\infty$, $\Delta\,(= T/n)\to 0$, $b\left(= b_{n,T}\right)\to 0$ such that

$$\frac{\bar{L}_X\,(T,x)}{b}\sqrt{\Delta\log\left(\frac{1}{\Delta}\right)} = o_{a.s.}\,(1)\,.$$

Assumption 4.4 ensures that the SDE (4.1) has a unique strong solution $X_t$ and that $X_t$ is recurrent. Assumption 4.5, together with Assumption 4.4, implies that the process $X_t$ is positive recurrent (or ergodic) and ensures the existence of a time-invariant distribution $P^0$ with density $f\,(x) = s\,(x)\,/\int_0^\infty s\,(x)\,dx$. In addition, if $X_0$ has distribution $P^0$, $X_t$ becomes strictly stationary. The square-root process by

Cox et al. (1985) and the inverse-Feller process by Ahn and Gao (1999), for instance, are known to have time-invariant distributions. Assumption 4.6 controls the rates of convergence or divergence of the sequences used in the asymptotic results for general and positive recurrent cases. $\bar{L}_X(T, x)$ is a normalized measure of the time spent by $X_t$ in the vicinity of a generic point $x$. When $X_t$ is strictly stationary, $\bar{L}_X(T, x)/T \overset{a.s.}{\to} f(x)$, where $f(x)$ is the time-invariant marginal density of $X_t$. In contrast, for a general recurrent diffusion, $\bar{L}_X(T, x)$ diverges to infinity at a rate no faster than $T$.

### 4.3.2.3  Convergence Properties of the Drift and Diffusion Estimators

The theorem below states asymptotic properties of the drift and diffusion estimators for general and positive recurrent cases separately.

**Theorem 4.3** **(i)** **(General Recurrent Case)** (Gospodinov and Hirukawa 2012, Theorem 1)
*Suppose that Assumptions 4.4 and 4.6 hold. Also let*

$$\widehat{\bar{L}}_X(T, x, b) := \Delta \sum_{i=1}^{n} K_{G(x,b)}(X_{i\Delta})$$

*be a (strongly) consistent estimate of $\bar{L}_X(T, x)$ for a fixed $T$.*
    *(a)* **(Drift Estimator)** *If $b^{5/2}\bar{L}_X(T, x) = O_{a.s.}(1)$, then, as $n, T \to \infty$,*

$$\sqrt{b^{1/2}\widehat{\bar{L}}_X(T, x, b)}\left\{\widehat{\mu}_b(x) - \mu(x) - B_\mu^R(x)b\right\} \overset{d}{\to} N\left(0, \frac{\sigma^2(x)}{2\sqrt{\pi}\sqrt{x}}\right)$$

*for $x/b \to \infty$, and*

$$\sqrt{b\widehat{\bar{L}}_X(T, x, b)}\left\{\widehat{\mu}_b(x) - \mu(x)\right\} \overset{d}{\to} N\left(0, \frac{\Gamma(2\kappa + 1)\sigma^2(x)}{2^{2\kappa+1}\Gamma^2(\kappa + 1)}\right)$$

*for $x/b \to \kappa \in (0, \infty)$, where*

$$B_\mu^R(x) = \left\{1 + x\frac{s^{(1)}(x)}{s(x)}\right\}\mu^{(1)}(x) + \frac{x}{2}\mu^{(2)}(x).$$

    *(b)* **(Diffusion Estimator)** *If $b^{5/2}\bar{L}_X(T, x)/\Delta = O_{a.s.}(1)$, then, as $n, T \to \infty$,*

$$\sqrt{\frac{b^{1/2}\widehat{\bar{L}}_X(T, x, b)}{\Delta}}\left\{\widehat{\sigma}_b^2(x) - \sigma^2(x) - B_{\sigma^2}^R(x)b\right\} \overset{d}{\to} N\left(0, \frac{\sigma^4(x)}{\sqrt{\pi}\sqrt{x}}\right)$$

*for $x/b \to \infty$, and*

$$\sqrt{\frac{b\widehat{\overline{L}}_X\left(T,x,b\right)}{\Delta}}\left\{\widehat{\sigma}_b^2(x)-\sigma^2\left(x\right)\right\}\overset{d}{\to}N\left(0,\frac{\Gamma\left(2\kappa+1\right)\sigma^4\left(x\right)}{2^{2\kappa}\Gamma^2\left(\kappa+1\right)}\right)$$

*for $x/b\to\kappa$, where*

$$B_{\sigma^2}^R\left(x\right)=\left\{1+x\frac{s^{(1)}\left(x\right)}{s\left(x\right)}\right\}\left\{\sigma^2\left(x\right)\right\}^{(1)}+\frac{x}{2}\left\{\sigma^2\left(x\right)\right\}^{(2)}.$$

(ii) (**Positive Recurrent Case**) (Gospodinov and Hirukawa 2012, Corollary 1)
   *Suppose that in addition to Assumptions 4.4 and 4.6, Assumption 4.5 holds.*
   (a) (**Drift Estimator**) *If $b\asymp T^{-2/5}$, then, as $n,T\to\infty$,*

$$\sqrt{Tb^{1/2}}\left\{\widehat{\mu}_b(x)-\mu(x)-B_\mu^S\left(x\right)b\right\}\overset{d}{\to}N\left(0,\frac{\sigma^2\left(x\right)}{2\sqrt{\pi}\sqrt{x}f\left(x\right)}\right)$$

*for $x/b\to\infty$, and*

$$\sqrt{Tb}\left\{\widehat{\mu}_b(x)-\mu(x)\right\}\overset{d}{\to}N\left(0,\frac{\Gamma\left(2\kappa+1\right)\sigma^2\left(x\right)}{2^{2\kappa+1}\Gamma^2\left(\kappa+1\right)f\left(x\right)}\right)$$

*for $x/b\to\kappa$, where*

$$B_\mu^S\left(x\right)=\left\{1+x\frac{f^{(1)}\left(x\right)}{f\left(x\right)}\right\}\mu^{(1)}(x)+\frac{x}{2}\mu^{(2)}(x).$$

(b) (**Diffusion Estimator**) *If $b\asymp n^{-2/5}$, then, as $n,T\to\infty$,*

$$\sqrt{nb^{1/2}}\left\{\widehat{\sigma}_b^2(x)-\sigma^2\left(x\right)-B_{\sigma^2}^S\left(x\right)b\right\}\overset{d}{\to}N\left(0,\frac{\sigma^4\left(x\right)}{\sqrt{\pi}\sqrt{x}f\left(x\right)}\right)$$

*for $x/b\to\infty$, and*

$$\sqrt{nb}\left\{\widehat{\sigma}_b^2(x)-\sigma^2\left(x\right)\right\}\overset{d}{\to}N\left(0,\frac{\Gamma\left(2\kappa+1\right)\sigma^4\left(x\right)}{2^{2\kappa}\Gamma^2\left(\kappa+1\right)f\left(x\right)}\right)$$

*for $x/b\to\kappa$, where*

$$B_{\sigma^2}^S\left(x\right)=\left\{1+x\frac{f^{(1)}\left(x\right)}{f\left(x\right)}\right\}\left\{\sigma^2\left(x\right)\right\}^{(1)}+\frac{x}{2}\left\{\sigma^2\left(x\right)\right\}^{(2)}.$$

In each part of Theorem 4.3, the rate on $b$ balances orders of magnitude in squared bias and variance for interior $x$. This rate undersmooths the curve for boundary $x$, and as a consequence, the $O\left(b\right)$ leading bias term becomes asymptotically negligible. The theorem also suggests that $\widehat{\mu}_b(x)$ has a slower convergence rate than $\widehat{\sigma}_b^2(x)$, regardless of the position of $x$. Therefore, if the sample size is not sufficiently large, it is much harder to estimate the drift accurately, especially for design points in the region of high values where the data are sparse. It also follows that a longer smoothing parameter is required to estimate the drift, as documented in Chapman and Pearson (2000, p. 367).

### 4.3.3 Additional Remarks

#### 4.3.3.1 Finite-Sample Properties of the Drift and Diffusion Estimators

Gospodinov and Hirukawa (2012) conduct Monte Carlo simulations based on the square-root process by Cox et al. (1985). The process is convenient because the transition and marginal densities are known and the prices of a zero-coupon bond and a call option written on the bond that are implied by the process have closed-form expressions.

The process is known to have a linear drift. Figure 2 of Gospodinov and Hirukawa (2012) reveals that the drift estimator using the G kernel captures linearity well and is practically unbiased. On the other hand, the drift estimator using the Gaussian kernel exhibits a downward bias over the region of high interest rate levels.

In addition, the bond and derivative prices based on the drift and diffusion estimators using the G kernel are less biased than those based on the Gaussian counterparts. The former enjoys much smaller variability and tighter confidence intervals than the latter. Furthermore, while the confidence interval of the call price based on the G kernel is roughly symmetric around the median estimate, that of the Gaussian-based call price appears to be highly asymmetric (with long right tail).

#### 4.3.3.2 A Few Words on Linearity in the Drift Function

There is still no consensus on the presence of statistically significant nonlinearity in the drift function of the US short-term interest rates. For instance, Chapman and Pearson (2000) argue that the nonlinearity in the drift of the spot rate at high values of interest rates documented by Stanton (1997) could be spurious due to the poor finite-sample properties of Stanton's (1997) estimator. Fan and Zhang (2003) conclude that there is little evidence against linearity in the short-rate drift function. In contrast, Arapis and Gao (2006) report that their specification testing strongly rejects the linearity of the short rate drift at both daily and monthly frequency.

Gospodinov and Hirukawa (2012) shed some light on this problem. Based on the bootstrap-based inference, Gospodinov and Hirukawa (2012) observe some mild (but statistically insignificant) nonlinearity in the drift function of US risk-free rates. However, their drift estimate at higher interest rate levels has much smaller curvature than the one originally reported by Stanton (1997).

## 4.4  Smoothing Parameter Selection

As in the cases of symmetric kernels, the CV and generalized cross-validation ("GCV") methods may be applied to implement asymmetric kernel regression estimation. For instance, Shi and Song (2016, p. 3495) define the GCV criterion for the NW estimator using the G kernel as

$$GCV\,(b) = \frac{n \sum_{i=1}^{n} \left( Y_i - \sum_{k=1}^{n} w_{ik} Y_k \right)^2}{\left( n - \sum_{i=1}^{n} w_{ii} \right)^2},$$

where

$$w_{ik} = \frac{K_{G(X_i,b)}\,(X_k)}{\sum_{\ell=1}^{n} K_{G(X_i,b)}\,(X_\ell)}, \; i, k = 1, \dots, n.$$

Below we discuss the CV method in the context of estimating scalar diffusion models. From the viewpoint of observations with serial dependence, Gospodinov and Hirukawa (2012) adopt the $h$-block CV method developed by Györfi et al. (1989, Chap. VI), Härdle and Vieu (1992) and Burman et al. (1994). For equally spaced observations $\{X_{i\Delta}\}_{i=1}^{n}$, let $\hat{m}_{-(i-h)\Delta:(i+h)\Delta}\,(X_{i\Delta})$ denote the estimate of the drift $\mu(X_{i\Delta})$ or the diffusion $\sigma^2(X_{i\Delta})$ from $n - (2h+1)$ observations

$$\left\{ X_\Delta, X_{2\Delta}, \dots, X_{(i-h-1)\Delta}, X_{(i+h+1)\Delta}, \dots, X_{n\Delta} (= X_T) \right\}.$$

The smoothing parameter can be selected by minimizing the $h$-block CV criterion

$$CV_h\,(b) = \sum_{i=h+1}^{n-h} \left\{ Y_{i\Delta} - \hat{m}_{-(i-h)\Delta:(i+h)\Delta}\,(X_{i\Delta}) \right\}^2,$$

where $Y_{i\Delta}$ is $\left( X_{(i+1)\Delta} - X_{i\Delta} \right)/\Delta$ (for drift estimation) or $\left( X_{(i+1)\Delta} - X_{i\Delta} \right)^2/\Delta$ (for diffusion estimation). Moreover, given some similarities between problems of choosing the block size $h$ and the bandwidth parameter in heteroskedasticity and autocorrelation consistent covariance estimation, Gospodinov and Hirukawa (2012) put $h = (\gamma n)^{1/4}$, where

$$\gamma := \frac{4\rho^2}{(1-\rho)^2(1+\rho)^2}$$

and $\rho \in (0, 1)$ is the coefficient when a first-order autoregressive model is fitted to $\{X_{i\Delta}\}_{i=1}^{n}$. In practice, $\rho$ is replaced by its least squares estimate. Lastly, observe that when $\gamma = 0$ (or equivalently $\rho = 0$), the $h$-block CV naturally reduces to the leave-one-out CV for serially uncorrelated data.

# References

Ahn, D.H., and B. Gao. 1999. A parametric nonlinear model of term structure dynamics. *Review of Financial Studies* 12: 721–762.

Arapis, M., and J. Gao. 2006. Empirical comparisons in short-term interest rate models using nonparametric methods. *Journal of Financial Econometrics* 4: 310–345.

Brownlees, C. T., F. Cipollini, and G. M. Gallo. 2012. Multiplicative error models. In *Handbook of Volatility Models and Their Applications*, ed. L. Bauwens, C. Hafner and S. Laurent, 223–247. Hoboken, NJ: John Wiley & Sons.

Burman, P., E. Chow, and D. Nolan. 1994. A cross-validatory method for dependent data. *Biometrika* 81: 351–358.

Chapman, D.A., and N.D. Pearson. 2000. Is the short rate drift actually nonlinear? *Journal of Finance* 55: 355–388.

Chen, S.X. 2000. Beta kernel smoothers for regression curves. *Statistica Sinica* 10: 73–91.

Chen, S.X. 2002. Local linear smoothers using asymmetric kernels. *Annals of the Institute of Statistical Mathematics* 54: 312–323.

Cox, J.C., J.E. Ingersoll Jr., and S.A. Ross. 1985. A theory of the term structure of interest rates. *Econometrica* 53: 385–408.

Engle, R.F. 2002. New frontiers for ARCH models. *Journal of Applied Econometrics* 17: 425–446.

Fan, J. 1993. Local linear regression smoothers and their minimax efficiencies. *Annals of Statistics* 21: 196–216.

Fan, J., and I. Gijbels. 1992. Variable bandwidth and local linear regression smoothers. *Annals of Statistics* 20: 2008–2036.

Fan, J., and C. Zhang. 2003. A reexamination of diffusion estimators with applications to financial model validation. *Journal of the American Statistical Association* 98: 118–134.

Gasser, T., and H.-G. Müller. 1979. Kernel estimation of regression functions. In *Smoothing Techniques for Curve Estimation: Proceedings of a Workshop Held in Heidelberg, April 2–4, 1979*, ed. T. Gasser and M. Rosenblatt, 23–68. Berlin: Springer-Verlag.

Gospodinov, N., and M. Hirukawa. 2012. Nonparametric estimation of scalar diffusion models of interest rates using asymmetric kernels. *Journal of Empirical Finance* 19: 595–609.

Gouriéroux, C., and J. Jasiak. 2006. Autoregressive gamma processes. *Journal of Forecasting* 25: 129–152.

Györfi, L., W. Härdle, P. Sarda, and P. Vieu. 1989. *Nonparametric Curve Estimation from Time Series*, Lecture Notes in Statistics. Berlin: Springer-Verlag.

Härdle, W., and P. Vieu. 1992. Kernel regression smoothing of time series. *Journal of Time Series Analysis* 13: 209–232.

Masry, E., and J. Fan. 1997. Local polynomial estimation of regression functions for mixing processes. *Scandinavian Journal of Statistics* 24: 165–179.

Nadaraya, É.A. 1964. On estimating regression. *Theory of Probability and Its Applications* 9: 141–142.

Seifert, B., and T. Gasser. 1996. Finite-sample variance of local polynomials: analysis and solutions. *Journal of the American Statistical Association* 91: 267–275.

Shi, J., and W. Song. 2016. Asymptotic results in gamma kernel regression. *Communications in Statistics – Theory and Methods* 45: 3489–3509.

Stanton, R. 1997. A nonparametric model of term structure dynamics and the market price of interest rate risk. *Journal of Finance* 52: 1973–2002.

Watson, G.S. 1964. Smooth regression analysis. *Sankhyā: The Indian Journal of Statistics, Series A* 26: 359–372.

# Chapter 5
# Specification Testing

There are only a few accomplishments on applying asymmetric kernels to specification testing. This chapter deals with three model specification tests. Each test has an application-driven flavor, and it is shown to be consistent, i.e., the power of the test approaches one as the sample size diverges. In particular, the test of discontinuity in densities in Sect. 5.3 extends an existing method and contains some new results. It is also worth emphasizing that asymmetric kernel tests perform well in finite samples despite relying simply on first-order asymptotic results. Therefore, assistance of size-adjusting devices such as bootstrapping appears to be unnecessary, unlike most of the smoothed tests employing conventional symmetric kernels. We conclude this chapter by discussing a method of choosing the smoothing parameter under the test-optimality criterion and providing technical proofs.

## 5.1 Test of a Parametric Form in Autoregressive Conditional Duration Models

### 5.1.1 Background

With the availability of high-frequency financial transaction data and the rapid advance in computing power, there is growing interest in applied microstructure research. In high-frequency financial econometrics, the timing of transactions is a key factor to understanding economic theory. For example, the time duration between market events has been found to have a deep impact on the behavior of market agents (e.g., traders and market makers) and on the intraday characteristics of the price process. Motivated by this feature, Engle and Russell (1998) propose a class of autoregressive conditional duration ("ACD") models to characterize the arrival time intervals between market events of interest such as the occurrence of a trade or a

© The Author(s) 2018
M. Hirukawa, *Asymmetric Kernel Smoothing*, JSS Research Series
in Statistics, https://doi.org/10.1007/978-981-10-5466-2_5

bid–ask quote. The main idea behind ACD modeling is a dynamic parameterization of the conditional expected duration given available public and private information.

Although a wide variety of ACD specifications have been proposed since the seminal work by Engle and Russell (1998), model evaluation has not received much attention up until recently. Fernandes and Grammig (2005) consider two nonparametric specification tests for the distribution of the standardized innovation (or the error term) in ACD models. Each test takes two steps. First, the conditional duration process is estimated by the quasi-maximum likelihood ("QML") method. Second, the functional-form misspecification of the distribution can be tested by gauging the closeness between parametric and nonparametric estimates of the baseline density or hazard function using the standardized residuals from the QML estimate ("QMLE"). The idea of comparing the parametric estimate which is consistent only under correct specification of the model with the nonparametric one which is consistent both under correct specification and misspecification of the model is closely related, for instance, to the density matching test for interest rate diffusion models by Aït-Sahalia (1996).

## 5.1.2 Specification Testing for the Distribution of the Standardized Innovation

### 5.1.2.1  Two Versions of the Test

Denote the duration by $z_i = t_i - t_{i-1}$, where $z_i$ is the time elapsed between events occurring at time $t_i$ and $t_{i-1}$. In ACD models, the duration is specified by $z_i = \psi_i \epsilon_i$, where $\psi_i = E(z_i | I_{i-1})$ is the conditional expected duration process given $I_{i-1}$, the information available at time $i - 1$, and $\epsilon_i$ ($\geq 0$) is the i.i.d. standardized innovation that is independent of $\psi_i$. Typically, $\psi_i$ is specified as a parametric form involving lagged $z_i$ and $\psi_i$, whereas the exponential, gamma, Weibull, and Burr distributions, for instance, are popularly chosen as the distribution of $\epsilon_i$.

The tests by Fernandes and Grammig (2005) are designed to detect misspecification of the distribution of $\epsilon_i$, under the assumption that the conditional expected duration process $\psi_i$ is correctly specified up to the parameter vector $\phi$. Let $f$ be the true pdf of the distribution of $\epsilon$. Also, a parametric family of $f$ is denoted by $\mathcal{P} = \{f(\cdot; \theta) : \theta \in \Theta\}$, where $\theta$ is the parameter vector and $\Theta$ is the parameter space. Fernandes and Grammig (2005) investigate the problem of testing the null hypothesis

$$H_0 : \exists \theta_0 \in \Theta \text{ such that } f(\cdot; \theta_0) = f(\cdot)$$

against the alternative

$$H_1 : f(\cdot) \notin \mathcal{P}.$$

Derivations of test statistics start from assuming that the true value of $\phi$ in the conditional expected duration process $\psi_i$ is known. It follows that the standardized innovation $\epsilon_i$ is observable, and thus the closeness between parametric and nonparametric densities $f(\cdot; \theta_0)$ and $f(\cdot)$ can be evaluated by the distance

$$\Phi_f := \int_0^\infty \{f(\epsilon; \theta) - f(\epsilon)\}^2 \, \mathbf{1}\{\epsilon \in \mathcal{S}\} \, dF(\epsilon),$$

where $\mathcal{S} \subseteq \mathbb{R}_+$ is a compact interval in which density estimation is stable so that $\mathbf{1}\{\epsilon \in \mathcal{S}\}$ serves as a trimming function. Using $n$ observations $\{\epsilon_i\}_{i=1}^n$ and replacing $\theta$, $f$, and $F$ with the MLE $\hat{\theta}$, the G density estimate $\hat{f}_G$, and the empirical measure $F_n$, respectively, we obtain the sample analog of $\Phi_f$ as

$$\Phi_{\hat{f}} := \frac{1}{n} \sum_{i=1}^n \left\{ f\left(\epsilon_i; \hat{\theta}\right) - \hat{f}_G(\epsilon_i) \right\}^2 \mathbf{1}\{\epsilon_i \in \mathcal{S}\}.$$

The specification test based on $\Phi_{\hat{f}}$ is labeled as the *D-test*.

It is also known that there is a one-to-one correspondence between the true pdf $f$ and the true hazard rate function $H_f(\cdot) := f(\cdot)/S(\cdot)$, where $S(\cdot) := 1 - F(\cdot)$ is the survival function. Accordingly, the null $H_0$ can be rewritten in the context of hazard-based testing as

$$H_0' : \exists \theta_0 \in \Theta \text{ such that } H_{\theta_0}(\cdot) = H_f(\cdot),$$

where $H_{\theta_0} := H(\cdot; \theta_0)$ is the hazard function implied by the parametric density $f(\cdot; \theta_0)$. Considering the sample analog to the distance between $H_{\theta_0}$ and $H_f$ as before, we have

$$\Lambda_{\hat{f}} := \frac{1}{n} \sum_{i=1}^n \left\{ H_{\hat{\theta}}(\epsilon_i) - H_{\hat{f}_G}(\epsilon_i) \right\}^2 \mathbf{1}\{\epsilon_i \in \mathcal{S}\},$$

where $H_{\hat{\theta}} := H\left(\cdot; \hat{\theta}\right)$ is the parametric estimate of the baseline hazard function with the MLE $\hat{\theta}$ plugged in and $H_{\hat{f}_G}$ is the nonparametric baseline hazard estimate using the G kernel. The specification test based on $\Lambda_{\hat{f}}$ is referred to as the *H-test*.

### 5.1.2.2 Convergence Properties of the D- and H-Tests

After providing a set of regularity conditions, we deliver convergence properties of the D- and H-tests.

**Assumption 5.1** $\{\epsilon_i\}_{i=1}^n$ are *i.i.d.* random variables drawn from a univariate distribution with a pdf $f$ having support on $\mathbb{R}_+$.

**Assumption 5.2** The pdf $f$ is twice continuously differentiable, and first two derivatives of $f$ are bounded and square integrable on $\mathbb{R}_+$. In addition, $f$ is bounded away from zero on the compact interval $\mathcal{S}$.

**Assumption 5.3** The smoothing parameter $b$ ($= b_n > 0$) satisfies $b \asymp n^{-q}$ for some constant $q \in (2/5, 1)$.

**Assumption 5.4** The parameter space $\Theta$ is a compact subset of $\mathbb{R}^p$. Moreover, there is a neighborhood $\mathcal{N}$ around $\theta_0$ such that $\zeta(\cdot; \theta)$, which is either $f(\cdot; \theta)$ or $H(\cdot; \theta)$, is twice continuously differentiable with respect to $\theta$ with uniformly bounded second-order partial derivatives and the matrix $E\left[\{\partial \zeta(\cdot; \theta) / \partial \theta\}\{\partial \zeta(\cdot; \theta) / \partial \theta\}^\top\right]$ is of full rank.

**Theorem 5.1** (Fernandes and Grammig 2005, Propositions 2 and 6)
   *Suppose that Assumptions 5.1–5.4 hold.*

(i) *Under $H_0$, as $n \to \infty$, the statistic for the D-test*

$$T_n^D := \frac{nb^{1/4}\Phi_{\hat{f}} - b^{-1/4}\hat{\delta}_G}{\hat{\sigma}_G} \xrightarrow{d} N(0, 1),$$

*where $\hat{\delta}_G$ and $\hat{\sigma}_G^2$ are consistent estimates of*

$$\delta_G := \frac{1}{2\sqrt{\pi}} E\left[\frac{f(\epsilon)}{\sqrt{\epsilon}}\mathbf{1}\{\epsilon \in \mathcal{S}\}\right] \text{ and } \sigma_G^2 := \frac{1}{2\sqrt{\pi}} E\left[\frac{\{f(\epsilon)\}^3}{\sqrt{\epsilon}}\mathbf{1}\{\epsilon \in \mathcal{S}\}\right],$$

*respectively.*

(ii) *Under $H_0'$, as $n \to \infty$, the statistic for the H-test*

$$T_n^H := \frac{nb^{1/4}\Lambda_{\hat{f}} - b^{-1/4}\hat{\lambda}_G}{\hat{\varsigma}_G} \xrightarrow{d} N(0, 1),$$

*where $\hat{\lambda}_G$ and $\hat{\varsigma}_G^2$ are consistent estimates of*

$$\lambda_G := \frac{1}{2\sqrt{\pi}} E\left[\frac{H_f(\epsilon)}{\sqrt{\epsilon}S(\epsilon)}\mathbf{1}\{\epsilon \in \mathcal{S}\}\right] \text{ and } \varsigma_G^2 := \frac{1}{2\sqrt{\pi}} E\left[\frac{\{H_f(\epsilon)\}^3}{\sqrt{\epsilon}S(\epsilon)}\mathbf{1}\{\epsilon \in \mathcal{S}\}\right],$$

*respectively.*

As $\hat{\delta}_G$, $\hat{\sigma}_G^2$, $\hat{\lambda}_G$, *and* $\hat{\varsigma}_G^2$, Fernandes and Grammig (2005) suggest

$$\hat{\delta}_G = \frac{1}{2\sqrt{\pi}} \frac{1}{n} \sum_{i=1}^{n} \frac{\hat{f}_G(\epsilon_i)}{\sqrt{\epsilon_i}} \mathbf{1}\{\epsilon_i \in \mathcal{S}\},$$

$$\hat{\sigma}_G^2 = \frac{1}{2\sqrt{\pi}} \frac{1}{n} \sum_{i=1}^{n} \frac{\left\{\hat{f}_G(\epsilon_i)\right\}^3}{\sqrt{\epsilon_i}} \mathbf{1}\{\epsilon_i \in \mathcal{S}\},$$

$$\hat{\lambda}_G = \frac{1}{2\sqrt{\pi}} \frac{1}{n} \sum_{i=1}^{n} \frac{\hat{f}_G(\epsilon_i)}{\sqrt{\epsilon_i}\left\{1 - \hat{F}_G(\epsilon_i)\right\}^2} \mathbf{1}\{\epsilon_i \in \mathcal{S}\}, \text{ and}$$

$$\hat{\varsigma}_G^2 = \frac{1}{2\sqrt{\pi}} \frac{1}{n} \sum_{i=1}^{n} \frac{\left\{\hat{f}_G(\epsilon_i)\right\}^3}{\sqrt{\epsilon_i}\left\{1 - \hat{F}_G(\epsilon_i)\right\}^4} \mathbf{1}\{\epsilon_i \in \mathcal{S}\},$$

where $\hat{F}_G(\epsilon) := \int_0^\epsilon \hat{f}_G(y)\,dy$ is the cdf estimate based on the G density estimate.

### 5.1.3 Additional Remarks

#### 5.1.3.1 Asymptotic Local Power and Consistency

Proposition 3 of Fernandes and Grammig (2005) demonstrates asymptotic local power of the D-test against the sequence of Pitman local alternatives $H_{1n} : f(\epsilon; \boldsymbol{\theta}) = f(\epsilon) + r_n h(\epsilon)$ for $r_n \asymp n^{-1/2} b^{-1/8}$ and $h(\epsilon)$ such that $\int_0^\infty h(\epsilon)\,dF(\epsilon) = 0$ and $\int_{\mathcal{S}} h^2(\epsilon)\,dF(\epsilon) \in (0, \infty)$. It is also straightforward to demonstrate that the power approaches one against the local alternatives with $r_n$ satisfying $r_n + 1/\left(n^{1/2} b^{1/8} r_n\right) \to 0$ as $n \to \infty$, as well as for the fixed alternative $H_1 : f(\epsilon; \boldsymbol{\theta}) = f(\epsilon) + h(\epsilon)$. Similar results apply for the H-test.

#### 5.1.3.2 Two-Step Specification Testing

The argument so far has been constructed under the assumption the true value of $\phi$ in the conditional expected duration process $\psi_i$ is known. In reality, however, the parameter $\phi$ is unknown. To make the D- and H-tests fully operational, we take a two-step procedure. In the first step, $\phi$ is estimated by QML. Denoting the QMLE by $\hat{\phi}$, we obtain the standardized residuals $\{\hat{\epsilon}_i\}_{i=1}^n = \left\{z_i/\hat{\psi}_i\right\}_{i=1}^n$, where $\hat{\psi}_i$ is the estimated conditional expected duration process with $\hat{\phi}$ plugged in. Then, in the second step, we proceed to either the D- or H-test using the standardized residuals. Proposition 9 of Fernandes and Grammig (2005) demonstrate that convergence properties of the D- and H-tests are unaffected even when $\phi$ is replaced by its $\sqrt{n}$-consistent estimate.

### 5.1.3.3  Finite-Sample Properties of Test Statistics

Fernandes and Grammig (2005) also consider the D- and H-tests with the density estimate using a standard symmetric kernel plugged in. Monte Carlo results indicate substantial size distortions of the tests using the symmetric kernel, which reflects that omitted terms in the first-order asymptotics on the test statistics are actually nonnegligible in their finite-sample distributions. As a consequence, Fernandes and Grammig (2005) turn to bootstrapping the standardized residuals for size correction. In contrast, the tests using the G kernel perform well in terms of both size and power. First-order asymptotic results on the test statistics appear to be well transmitted to their finite-sample distributions, which make bootstrapping unnecessary.

## 5.2  Test of Symmetry in Densities

### 5.2.1  Background

Symmetry and conditional symmetry play a key role in numerous fields of economics and finance. In econometrics, for example, conditional symmetry in the distribution of the disturbance is often a key regularity condition for regression analyses including adaptive estimation and robust regression estimation. In finance, the mean-variance analysis is consistent with investors' portfolio decision making if and only if asset returns are elliptically distributed.

In view of the importance in the existence of symmetry, Fernandes et al. (2015) propose the split-sample symmetry test ("SSST") smoothed by the G kernel. Later, Hirukawa and Sakudo (2016) ameliorate the SSST by combining it with the MG and NM kernels. While superior finite-sample performance of the MG kernel has been reported in the literature, the NM kernel is also anticipated to have an advantage when applied to the SSST. It is known that finite-sample performance of a kernel density estimator depends on proximity in shape between the underlying density and the kernel chosen. As illustrated in Fig. 1.3, the NM kernel collapses to the half-normal pdf when smoothing is made at the origin, and the shape of the density is likely to be close to those on the positive side of single-peaked symmetric distributions.

### 5.2.2  Asymmetric Kernel-Based Testing for Symmetry in Densities

#### Case (i): When Two Subsamples Have Equal Sample Sizes

Suppose, without loss of generality, that we are interested in testing symmetry about zero of the distribution of a random variable $U \in \mathbb{R}$. If $U$ has a pdf, then under the null of symmetry, its shapes on positive and negative sides of the entire real line $\mathbb{R}$ must be mirror images each other. Therefore, whether symmetry holds can be tested by gauging closeness between the density estimates on positive and negative sides from positive and absolute values of negative observations, respectively. Let $f$ and $g$ be the pdfs to the right and left from the origin, respectively. Given $f$ and $g$, the null and alternative hypotheses can be expressed as

$$H_0 : f(u) = g(u) \text{ for almost all } u \in \mathbb{R}_+, \text{ and}$$
$$H_1 : f(u) \neq g(u) \text{ on a set of positive measure in } \mathbb{R}_+,$$

respectively. A natural test statistic may be built on the ISE between $f$ and $g$

$$I = \int_0^\infty \{f(u) - g(u)\}^2 \, du$$
$$= \int_0^\infty \{f(u) - g(u)\} \, dF(u) - \int_0^\infty \{f(u) - g(u)\} \, dG(u),$$

where $F$ and $G$ are cdfs corresponding to $f$ and $g$, respectively.

To construct a sample analog to $I$, we split the random sample $\{U_i\}_{i=1}^N$ into two sub-samples, namely $\{X_i\}_{i=1}^{n_1} := \{U_i : U_i \geq 0\}_{i=1}^{n_1}$ and $\{Y_i\}_{i=1}^{n_2} := \{-U_i : U_i < 0\}_{i=1}^{n_2}$, where $N = n_1 + n_2$. Then, we employ one of the G, MG, and NM kernels to estimate $f$ and $g$, respectively, as

$$\hat{f}_j(u) = \frac{1}{n_1} \sum_{i=1}^{n_1} K_{j(u,b)}(X_i) \text{ and}$$

$$\hat{g}_j(u) = \frac{1}{n_2} \sum_{i=1}^{n_2} K_{j(u,b)}(Y_i), \quad j \in \{G, MG, NM\}.$$

In addition, because $n_1 \sim n_2$ under $H_0$, without loss of generality and for ease of exposition, we assume that $N$ is even and that $n := n_1 = n_2 = N/2$. Substituting $(\hat{f}_j, \hat{g}_j)$ in places of $(f, g)$ and replacing $(F, G)$ with their empirical measures $(F_{n_1}, G_{n_2})$ yield the sample analog to $I$ as

$$\bar{I}_{n,j} = \frac{1}{n} \sum_{i=1}^n \left\{ \hat{f}_j(X_i) + \hat{g}_j(Y_i) - \hat{g}_j(X_i) - \hat{f}_j(Y_i) \right\}$$

$$= \sum_{i=1}^n \frac{1}{n^2} \left\{ K_{j(X_i,b)}(X_i) + K_{j(Y_i,b)}(Y_i) - K_{j(Y_i,b)}(X_i) - K_{j(X_i,b)}(Y_i) \right\}$$

$$+ \sum_{k=1}^{n} \sum_{i=1, i \neq k}^{n} \frac{1}{n^2} \left\{ K_{j(X_k, b)}(X_i) + K_{j(Y_k, b)}(Y_i) - K_{j(Y_k, b)}(X_i) - K_{j(X_k, b)}(Y_i) \right\}$$

$$= I_{1n, j} + I_{n, j}, \quad j \in \{G, MG, NM\}.$$

Although we could use $\bar{I}_{n, j}$ itself as the test statistic, $I_{1n, j}$ is likely to the source of size distortions in finite samples in that its probability limit plays a role in a nonvanishing center term of the asymptotic null distribution. Instead of subtracting $I_{1n, j}$ from $\bar{I}_{n, j}$ like the D- and H-tests in Sect. 5.1, we build the test statistic solely on $I_{n, j}$. Observe that $I_{n, j}$ can be reformulated as

$$I_{n, j} := \sum_{1 \leq i < k \leq n} \Phi_{n, j}(Z_i, Z_k) := \sum_{1 \leq i < k \leq n} \frac{1}{n^2} \left\{ \phi_{n, j}(Z_i, Z_k) + \phi_{n, j}(Z_k, Z_i) \right\},$$

where $Z_i := (X_i, Y_i)$ and

$$\phi_{n, j}(Z_i, Z_k) := K_{j(X_k, b)}(X_i) + K_{j(Y_k, b)}(Y_i) - K_{j(Y_k, b)}(X_i) - K_{j(X_k, b)}(Y_i).$$

It can be found that $I_{n, j}$ is a degenerate $U$-statistic, because $\Phi_{n, j}(Z_i, Z_k)$ is symmetric between $Z_i$ and $Z_k$ and $E\left\{ \Phi_{n, j}(Z_i, Z_k) \middle| Z_i \right\} = 0$ almost surely under $H_0$. Therefore, a martingale central limit theorem applies to $I_{n, j}$.

After presenting a set of regularity conditions, we document convergence results of $I_{n, j}$.

**Assumption 5.5**  Two random samples $\{X_i\}_{i=1}^{n_1}$ and $\{Y_i\}_{i=1}^{n_2}$ are drawn independently from univariate distributions that have pdfs $f$ and $g$ with support on $\mathbb{R}_+$, respectively.

**Assumption 5.6**  $f$ and $g$ are twice continuously differentiable on $\mathbb{R}_+$, and $E\left| X f^{(2)}(X) \right|^2$, $E\left| X^2 f^{(2)}(X) g^{(2)}(X) \right|$, $E\left| Y^2 f^{(2)}(Y) g^{(2)}(Y) \right|$, $E\left| Y g^{(2)}(Y) \right|^2 < \infty$.

**Theorem 5.2**  (Fernandes et al. 2015, Proposition 1; Hirukawa and Sakudo 2016, Theorem 1)
*Suppose that Assumptions 5.5 and 5.6 and $n_1 = n_2 = n$ hold and that the smoothing parameter $b (= b_n)$ satisfies $b + (nb)^{-1} \to 0$ as $n \to \infty$.*

(i)  *Under $H_0$, $nb^{1/4} I_{n, j} \xrightarrow{d} N\left(0, \sigma_j^2\right)$, $j \in \{G, MG, NM\}$ as $n \to \infty$, where*

$$\sigma_j^2 = 2v_j E\left[ X^{-1/2} \{ f(X) + g(X) \} + Y^{-1/2} \{ f(Y) + g(Y) \} \right]$$

*and $v_j$ is the kernel-specific constant given in (2.4).*

(ii)  *A consistent estimator of $\sigma_j^2$ is given by*

$$\hat{\sigma}_j^2 = 2v_j \frac{1}{n} \sum_{i=1}^{n} \left[ X_i^{-1/2} \left\{ \hat{f}_j(X_i) + \hat{g}_j(X_i) \right\} + Y_i^{-1/2} \left\{ \hat{f}_j(Y_i) + \hat{g}_j(Y_i) \right\} \right].$$

The variance estimator $\hat{\sigma}_j^2$ is consistent both under $H_0$ and $H_1$. Moreover, $\sigma_j^2$ reduces to

$$\sigma_j^2 = 8v_j E\left\{X^{-1/2} f(X)\right\}$$

under $H_0$, and thus Fernandes et al. (2015) exclusively employ its sample analog as the variance estimator. The theorem implies that the test statistic is

$$T_{n,j} := nb^{1/4}\frac{I_{n,j}}{\hat{\sigma}_j} \xrightarrow{d} N(0,1) \text{ under } H_0,$$

as $n \to \infty$. Clearly, the SSST is a one-sided test that rejects $H_0$ in favor of $H_1$ if $T_{n,j} > z_\alpha$, where $z_\alpha$ is the upper $\alpha$-percentile of $N(0,1)$.

**Case (ii): When Two Subsamples Have Unequal Sample Sizes**

When the sample sizes of two subsamples $\{X_i\}_{i=1}^{n_1}$ and $\{Y_i\}_{i=1}^{n_2}$ differ, i.e., $n_1 \neq n_2$ is the case, $I_{n,j}$ can be rewritten as

$$
\begin{aligned}
I_{n_1,n_2,j} =& \sum_{k=1}^{n_1}\sum_{i=1,i\neq k}^{n_1}\frac{1}{n_1^2}K_{j(X_k,b)}(X_i) + \sum_{k=1}^{n_2}\sum_{i=1,i\neq k}^{n_2}\frac{1}{n_2^2}K_{j(Y_k,b)}(Y_i) \\
&- \sum_{k=1}^{n_2}\sum_{i=1,i\neq k}^{n_1}\frac{1}{n_1 n_2}K_{j(Y_k,b)}(X_i) - \sum_{k=1}^{n_1}\sum_{i=1,i\neq k}^{n_2}\frac{1}{n_1 n_2}K_{j(X_k,b)}(Y_i) \quad (5.1)
\end{aligned}
$$

for $j \in \{G, MG, NM\}$. Following Fan and Ullah (1999), we deliver convergence results under the assumption that two sample sizes $n_1$ and $n_2$ diverge at the same rate.

**Theorem 5.3** (Fernandes et al. 2015, Proposition 2; Hirukawa and Sakudo 2016, Theorem 2)

*Suppose that Assumptions 5.5 and 5.6 and $n_1/n_2 \to \lambda$ for some constant $\lambda \in (0, \infty)$ hold and that the smoothing parameter $b\left(= b_{n_1}\right)$ satisfies $b + (n_1 b)^{-1} \to 0$ as $n_1 \to \infty$.*

(i) *Under $H_0$, $n_1 b^{1/4} I_{n_1,n_2,j} \xrightarrow{d} N\left(0, \sigma_{\lambda,j}^2\right)$, $j \in \{G, MG, NM\}$ as $n_1 \to \infty$, where*

$$
\begin{aligned}
\sigma_{\lambda,j}^2 =& 2v_j\Big[E\left\{X^{-1/2}f(X)\right\} + \lambda E\left\{X^{-1/2}g(X)\right\} \\
&+ \lambda E\left\{Y^{-1/2}f(Y)\right\} + \lambda^2 E\left\{Y^{-1/2}g(Y)\right\}\Big],
\end{aligned}
$$

*and $v_j$ is again given in (2.4).*

(ii) *A consistent estimator of $\sigma_{\lambda,j}^2$ is given by*

$$\hat{\sigma}^2_{\lambda,j} = 2v_j \left\{ \frac{1}{n_1} \sum_{i=1}^{n_1} X_i^{-1/2} \hat{f}_j(X_i) + \left(\frac{n_1}{n_2}\right) \frac{1}{n_1} \sum_{i=1}^{n_1} X_i^{-1/2} \hat{g}_j(X_i) \right.$$

$$\left. + \left(\frac{n_1}{n_2}\right) \frac{1}{n_2} \sum_{i=1}^{n_2} Y_i^{-1/2} \hat{f}_j(Y_i) + \left(\frac{n_1}{n_2}\right)^2 \frac{1}{n_2} \sum_{i=1}^{n_2} Y_i^{-1/2} \hat{g}_j(Y_i) \right\}. \quad (5.2)$$

Theorem 5.3 presents the limiting null distribution of $n_1 b^{1/4} I_{n_1,n_2,j}$, which implies that the test statistic in this case takes the form of

$$T_{n_1,n_2,j} := n_1 b^{1/4} \frac{I_{n_1,n_2,j}}{\hat{\sigma}_{\lambda,j}} \xrightarrow{d} N(0,1) \text{ under } H_0,$$

as $n_1 \to \infty$. Again the variance estimator $\hat{\sigma}^2_{\lambda,j}$ given in Theorem 5.3 is consistent both under $H_0$ and $H_1$, and $\sigma^2_{\lambda,j}$ reduces to

$$\sigma^2_{\lambda,j} = 2(1+\lambda)^2 v_j E\left\{X^{-1/2} f(X)\right\}$$

under $H_0$. Finally, when $n_1 = n_2 = n$ so that $\lambda \equiv 1$, Theorem 5.3 collapses to Theorem 5.2.

### 5.2.3   Additional Remarks

#### 5.2.3.1   A Comparison with the Test of Symmetry Using Symmetric Kernels

Applying the idea of two-sample goodness-of-fit tests to the symmetry test is not new. Ahmad and Li (1997) and Fan and Ullah (1999) have also studied the symmetry test based on closeness of two density estimates measured by the ISE. They estimate densities using two samples, namely the original entire sample $\{X_i\}_{i=1}^N := \{U_i\}_{i=1}^N$ and the one obtained by flipping the sign of each observation $\{Y_i\}_{i=1}^N := \{-U_i\}_{i=1}^N$ in our notations. Because each of $X$ and $Y$ has support on $(-\infty, \infty)$ by construction, a standard symmetric kernel is employed for density estimation. Notice that while these tests require continuity of density derivatives at the origin, the SSST does not.

#### 5.2.3.2   Two-Sample Test of Equality in Two Densities

If $X$ and $Y$ are taken from two different distributions with support on $\mathbb{R}_+$, then $I_{n,j}$ (or $I_{n_1,n_2,j}$) can be viewed as a pure two-sample goodness-of-fit test. It can be immediately applied to the testing for equality of two unknown distributions of nonnegative economic and financial variables such as incomes, wages, short-term interest rates, and insurance claims.

### 5.2.3.3 Asymptotic Local Power and Consistency

Proposition 4 of Fernandes et al. (2015) establishes nontrivial power of the SSST against the sequence of Pitman local alternatives $H_{1n} : g(u) = f(u) + r_n h(u)$ for $r_n \asymp n^{-1/2}b^{-1/8}$ and $h(u)$ such that $\int_0^\infty h(u)\,du = 0$ and $\int_0^\infty h^2(u)\,du \in (0, \infty)$. The test is also consistent against the local alternatives with $r_n$ satisfying $r_n + 1/\left(n^{1/2}b^{1/8}r_n\right) \to 0$ as $n \to \infty$, as well as for the fixed alternative $H_1 : g(u) = f(u) + h(u)$; see Propositions 1 (for Case (i)) and 2 (for Case (ii)) of Hirukawa and Sakudo (2016) for more details.

### 5.2.3.4 Finite-Sample Properties of the Test Statistic

Monte Carlo simulations in Fernandes et al. (2015) and Hirukawa and Sakudo (2016) indicate nice finite-sample performance of the SSST. The performance is confirmed even when the entire sample size $N$ is 50, despite a nonparametric convergence rate and a sample-splitting procedure. It should be again stressed that the good performance is based solely on first-order asymptotic results.

### 5.2.3.5 Test of Conditional Symmetry

Our focus so far has been on testing for unconditional symmetry. However, often $U$ is unobservable (e.g., the error term in a regression model) or the axis of symmetry is not zero (e.g., a nonzero mean or median). For these cases, the SSST must be extended to the test for conditional symmetry. Proposition 5 of Fernandes et al. (2015) and Theorem 3 of Hirukawa and Sakudo (2016) demonstrate that the SSST may be used for the test of conditional symmetry, as long as unknown parameters are replaced by their $\sqrt{n}$-consistent estimates and the smoothing parameter is chosen as $b \asymp n^{-q}$ for some $q \in (2/5, 1)$.

## 5.3 Test of Discontinuity in Densities

### 5.3.1 Background

Economic importance and interest in discontinuity of densities can be argued in the context of regression discontinuity designs ("RDD"). Local randomization of a continuous running variable is a key requirement for the validity of RDD; if the value of the running variable falls into the left and right of the cutoff strategically, then treatment effects are no longer point identified due to self-selection. Therefore, detection of discontinuity in the density of the running variable at the cutoff suggests evidence of such strategic behavior or manipulation in RDD.

Nonetheless, there are only a few estimation and testing procedures on discontinuity in densities available. Examples include the bin-based LL regression method by McCrary (2008), the empirical likelihood-based inference by Otsu et al. (2013), and the truncated kernel method built on the G kernel by Funke and Hirukawa (2017a). While all these previous works exclusively consider the case of a single cutoff that is suspected to be a discontinuity point, this section proposes a joint test of no jumps at prespecified multiple cutoffs in a density with support on $\mathbb{R}_+$. The test is a natural extension of the one by Funke and Hirukawa (2017a).

## 5.3.2  Joint Estimation and Testing on Discontinuity in Densities at Multiple Cutoffs

### 5.3.2.1  Estimation of the Jump Size via the Truncated Kernel Method

Suppose that for some small positive integer $L$ we suspect discontinuity of the pdf $f$ at $L$ prespecified points or cutoffs $c_1, \ldots, c_L$, where $(c_0 \equiv) \, 0 < c_1 < \cdots < c_L < \infty \, (\equiv c_{L+1})$. Throughout it is assumed that two adjacent cutoffs $c_\ell$ and $c_{\ell+1}$ for $\ell = 0, \ldots, L$ are so separate that there are a sufficient number of observations between them. Let

$$f_- (c_\ell) := \lim_{x \uparrow c_\ell} f(x) \text{ and } f_+ (c_\ell) := \lim_{x \downarrow c_\ell} f(x)$$

be the lower and upper limits of the pdf at $x = c_\ell$ for $\ell = 1, \ldots, L$, respectively. Then, the jump-size magnitude of the density at $c_\ell$ is

$$J(c_\ell) := f_+ (c_\ell) - f_- (c_\ell) .$$

To check whether $f$ is continuous at $L$ cutoffs $c_1, \ldots, c_L$ jointly, we first estimate $\mathbf{J}(\mathbf{c}) := (J(c_1), \ldots, J(c_L))^\top$ nonparametrically and then proceed to a hypothesis testing for the null of continuity of $f$ at all $c_1, \ldots, c_L$, i.e.,

$$H_0 : \mathbf{J}(\mathbf{c}) = \mathbf{0},$$

against the alternative of discontinuity of $f$ at some of $c_1, \ldots, c_L$, i.e.,

$$H_1 : \mathbf{J}(\mathbf{c}) \neq \mathbf{0}.$$

For notational conciseness, expressions such as "$f_\pm (c_\ell)$" are used throughout, whenever no confusions may occur.

To develop a consistent estimator of $\mathbf{J}(\mathbf{c})$, we generalize the truncated kernel method by Funke and Hirukawa (2017a) to the case of multiple cutoffs. The G kernel is split into $(L + 1)$ disjoint parts so that

$$K_{G(x,b)}(u) := \sum_{\ell=0}^{L} K_{G(x,b)}^{[c_\ell, c_{\ell+1})}(u) := \sum_{\ell=0}^{L} \frac{u^{x/b} \exp(-u/b)}{b^{x/b+1} \Gamma(x/b + 1)} \mathbf{1}\{u \in [c_\ell, c_{\ell+1})\}.$$

Because $K_{G(x,b)}^{[c_\ell, c_{\ell+1})}(u)$ is not a legitimate kernel function in the sense that

$$\int_0^\infty K_{G(x,b)}^{[c_\ell, c_{\ell+1})}(u)\,du = \frac{\gamma(x/b+1, c_{\ell+1}/b) - \gamma(x/b+1, c_\ell/b)}{\Gamma(x/b+1)} \in (0, 1),$$

we renormalize $K_{G(x,b)}^{[c_\ell, c_{\ell+1})}(u)$ to construct the truncated kernel

$$K_{G(x,b;c_\ell,c_{\ell+1})}(u) := \frac{\Gamma(x/b+1)}{\gamma(x/b+1, c_{\ell+1}/b) - \gamma(x/b+1, c_\ell/b)} K_{G(x,b)}^{[c_\ell, c_{\ell+1})}(u)$$

for $\ell = 0, \ldots, L$. It follows from $(c_0, c_{L+1}) = (0, \infty)$ that $\gamma(x/b+1, c_{\ell+1}/b) - \gamma(x/b+1, c_\ell/b)$ for $\ell = 0, L$ reduce to $\gamma(x/b+1, c_1/b)$ and $\Gamma(x/b+1, c_L/b)$, respectively.

Given the random sample $\{X_i\}_{i=1}^n$ drawn from a univariate distribution with a pdf $f$, we can estimate $J(c_\ell)$ for $\ell = 1, \ldots, L$ consistently as the difference between consistent estimates of $f_+(c_\ell)$ and $f_-(c_\ell)$. While

$$\hat{f}_+(c_\ell) = \frac{1}{n} \sum_{i=1}^n K_{G(x,b;c_\ell,c_{\ell+1})}(X_i)\big|_{x=c_\ell} = \frac{1}{n} \sum_{i=1}^n K_{G(c_\ell,b;c_\ell,c_{\ell+1})}(X_i) \text{ and}$$

$$\hat{f}_-(c_\ell) = \frac{1}{n} \sum_{i=1}^n K_{G(x,b;c_{\ell-1},c_\ell)}(X_i)\big|_{x=c_\ell} = \frac{1}{n} \sum_{i=1}^n K_{G(c_\ell,b;c_{\ell-1},c_\ell)}(X_i)$$

are consistent for $f_+(c_\ell)$ and $f_-(c_\ell)$, respectively, each estimator has inferior bias convergence due to one-sided smoothing as demonstrated in Proposition 1 of Funke and Hirukawa (2017a). Then, we define the consistent estimator of $J(c_\ell)$ as

$$\tilde{J}(c_\ell) := \tilde{f}_+(c_\ell) - \tilde{f}_-(c_\ell), \quad \ell = 1, \ldots, L,$$

where

$$\tilde{f}_\pm(c_\ell) := \left\{\hat{f}_{\pm,b}(c_\ell)\right\}^{1/(1-\delta^{1/2})} \left\{\hat{f}_{\pm,b/\delta}(c_\ell)\right\}^{-\delta^{1/2}/(1-\delta^{1/2})}$$

for some constant $\delta \in (0, 1)$ are the TS-MBC estimators of $f_\pm(c_\ell)$ in Sect. 3.2.2.1 and $\hat{f}_{\bullet,b}(x)$ and $\hat{f}_{\bullet,b/\delta}(x)$ signify the density estimators using smoothing parameters $b$ and $b/\delta$, respectively.

### 5.3.2.2    A Rationale for the Truncated Kernel Method

Some readers may wonder why we avoid estimating $f_-(c_\ell)$ and $f_+(c_\ell)$ using the B and/or G kernels. The reason is bias and variance convergences of the density estimates using these kernels when the cutoff $c_\ell$ is chosen as the boundary. In the vicinity of the cutoff, their bias convergences are usual $O(b)$, whereas the variance convergences slow down to $O\left(n^{-1}b^{-1}\right)$. Although the inferior rate is likely to adversely affect power properties of our test, it is hard (or even impossible) to improve the rate to usual $O\left(n^{-1}b^{-1/2}\right)$. In fact, in a closely related study, Fé (2014) leaves the inferior convergence rate of his RDD estimator smoothed by the G kernel as it is. In contrast, density estimates using the truncated kernels have the inferior bias convergence of $O\left(b^{1/2}\right)$ with the usual variance convergence of $O\left(n^{-1}b^{-1/2}\right)$ maintained. While the slow bias convergence potentially has a negative impact on size properties of the test, it can be improved to usual $O(b)$ with no additional conditions (e.g., extra smoothness in the density), by means of TS-MBC. A similar idea can be found in Guillamón et al. (1999), who adopt jackknife bias correction methods to improve the bias convergence of Bagai and Prakasa Rao's (1995) density estimator smoothed by a one-sided kernel.

### 5.3.2.3    Test Statistic for Discontinuity

To establish convergence results of $\widetilde{\mathbf{J}}(\mathbf{c})$, we assume a set of regularity conditions below.

**Assumption 5.7**  $\{X_i\}_{i=1}^n$ are *i.i.d.* random variables drawn from a univariate distribution with a pdf $f$ having support on $\mathbb{R}_+$.

**Assumption 5.8**  For each of $\ell = 1, \ldots, L$, there is a neighborhood $\mathcal{N}_\ell$ around $c_\ell$ such that the second-order derivative of the pdf $f$ is continuous on $\mathcal{N}_\ell \backslash \{c_\ell\}$. Also, let $f_-^{(p)}(c_\ell) := \lim_{x \uparrow c_\ell} d^p f(x)/dx^p$ and $f_+^{(p)}(c_\ell) := \lim_{x \downarrow c_\ell} d^p f(x)/dx^p$ for $p = 1, 2$. Then, $f_\pm(c_\ell) > 0$ and $\left| f_\pm^{(2)}(c_\ell) \right| < \infty$.

**Assumption 5.9**  The smoothing parameter $b(= b_n > 0)$ satisfies $b + \left(nb^{3/2}\right)^{-1} \to 0$ as $n \to \infty$.

**Theorem 5.4**  *Let* $\widetilde{\mathbf{J}}(\mathbf{c}) := \left( \tilde{J}(c_1), \ldots, \tilde{J}(c_L) \right)^\top$. *Also, suppose that Assumptions 5.7–5.9 hold.*

**(i)  (The Bias-Variance Trade-off)**
   *As $n \to \infty$,*

$$Bias\left\{\widetilde{\mathbf{J}}(\mathbf{c})\right\} = \mathbf{B}(\mathbf{c})b + o(b), \quad and$$

$$Var\left\{\widetilde{\mathbf{J}}(\mathbf{c})\right\} = \frac{1}{nb^{1/2}}\mathbf{V}(\mathbf{c}) + o\left(n^{-1}b^{-1/2}\right),$$

where $\mathbf{B}(\mathbf{c}) := (B(c_1), \ldots, B(c_L))^\top$, $\mathbf{V}(\mathbf{c}) := \mathrm{diag}\{V(c_1), \ldots, V(c_L)\}$,

$$
B(c_\ell) = \left(\frac{1}{\delta^{1/2}}\right)\left[\frac{c_\ell}{\pi}\left\{\frac{\left(f_+^{(1)}(c_\ell)\right)^2}{f_+(c_\ell)} - \frac{\left(f_-^{(1)}(c_\ell)\right)^2}{f_-(c_\ell)}\right\}\right.
$$
$$
\left. - \left\{\left(1 - \frac{4}{3\pi}\right)\left(f_+^{(1)}(c_\ell) - f_-^{(1)}(c_\ell)\right) + \frac{c_\ell}{2}\left(f_+^{(2)}(c_\ell) - f_-^{(2)}(c_\ell)\right)\right\}\right],
$$

$$
V(c_\ell) = v(\delta)\left\{\frac{f_+(c_\ell) + f_-(c_\ell)}{\sqrt{\pi}c_\ell^{1/2}}\right\}
$$

for $\ell = 1, \cdots, L$, and

$$
v(\delta) := \frac{\left(1 + \delta^{3/2}\right)(1 + \delta)^{1/2} - 2\sqrt{2}\delta}{(1 + \delta)^{1/2}\left(1 - \delta^{1/2}\right)^2}
$$

is monotonously increasing in $\delta \in (0, 1)$ with

$$
\lim_{\delta\downarrow 0} v(\delta) = 1 \text{ and } \lim_{\delta\uparrow 1} v(\delta) = \frac{11}{4}.
$$

**(ii) (Asymptotic Normality)**
  In addition, if $nb^{5/2} \to 0$ as $n \to \infty$, then

$$
\sqrt{nb^{1/2}}\left\{\widetilde{\mathbf{J}}(\mathbf{c}) - \mathbf{J}(\mathbf{c})\right\} \xrightarrow{d} N_L(\mathbf{0}, \mathbf{V}(\mathbf{c})).
$$

*Proof* See Sect. 5.5.1.
  Part (i) of the theorem suggests that the leading bias term $\mathbf{B}(\mathbf{c})\, b$ cancels out if $f$ has a continuous second-order derivative at each of $L$ cutoffs. It also follows from Part (ii) of the theorem that given a smoothing parameter $b \asymp n^{-q}$ for some constant $q \in (2/5, 2/3)$ (which fulfills three rate requirements for $b$, namely $b \to 0$, $nb^{3/2} \to \infty$ and $nb^{5/2} \to 0$ as $n \to \infty$) and $\widetilde{\mathbf{V}}(\mathbf{c})$, a consistent estimate of $\mathbf{V}(\mathbf{c})$, the test statistic is

$$
T(\mathbf{c}) = nb^{1/2}\widetilde{\mathbf{J}}(\mathbf{c})^\top\widetilde{\mathbf{V}}(\mathbf{c})^{-1}\widetilde{\mathbf{J}}(\mathbf{c})
$$
$$
= \sum_{\ell=1}^{L} nb^{1/2}\frac{\left\{\widetilde{J}(c_\ell)\right\}^2}{\widetilde{V}(c_\ell)} \xrightarrow{d} \chi^2(L) \text{ under } H_0 : \mathbf{J}(\mathbf{c}) = \mathbf{0},
$$

where $\chi^2(L)$ denotes the chi-squared distribution with $L$ degrees of freedom. The test rejects $H_0$ in favor of $H_1$ if $T(\mathbf{c}) > \chi_\alpha^2(L)$, where $\chi_\alpha^2(L)$ is the upper $\alpha$-percentile of $\chi^2(L)$. Observe that the test investigated in Funke and Hirukawa (2017a) corresponds to the case with $L = 1$.

#### 5.3.2.4    Asymptotic Local Power and Consistency of the Test

An argument like Proposition 3 of Funke and Hirukawa (2017a) immediately establishes asymptotic local power and consistency of the test. Specifically, the test has nontrivial power against the sequence of Pitman local alternatives $H_{1n} : \mathbf{J}_n\,(\mathbf{c}) = \mathbf{0} + r_n\mathbf{C}_1$ for $r_n \asymp n^{-1/2}b^{-1/4}$ and some constant vector $\|\mathbf{C}_1\| \in (0, \infty)$. The test is also consistent against the local alternatives with $r_n$ satisfying $r_n + 1/\left(n^{1/2}b^{1/4}r_n\right) \to 0$ as $n \to \infty$, as well as for the fixed alternative $H_1 : \mathbf{J}\,(\mathbf{c}) = \mathbf{C}_2$ for some constant vector $\|\mathbf{C}_2\| \in (0, \infty)$.

#### 5.3.2.5    Consistent Estimates of V (c)

There are a few candidates of $\widetilde{\mathbf{V}}\,(\mathbf{c})$. Replacing $f_\pm\,(c_\ell)$ in $V\,(c_\ell)$ with their consistent estimates $\hat{f}_\pm\,(c_\ell)$ immediately yields $\widetilde{\mathbf{V}}_1\,(\mathbf{c}) := \mathrm{diag}\left\{\tilde{V}_1\,(c_1),\ldots,\tilde{V}_1\,(c_L)\right\}$, where

$$\tilde{V}_1\,(c_\ell) := \upsilon\,(\delta)\left\{\frac{\hat{f}_+\,(c_\ell) + \hat{f}_-\,(c_\ell)}{\sqrt{\pi}c_\ell^{1/2}}\right\}.$$

Alternatively, taking into account that the G density estimator at $c_\ell$

$$\hat{f}_G\,(c_\ell) := \frac{1}{n}\sum_{i=1}^{n} K_{G(x,b)}\,(X_i)\Big|_{x=c_\ell} = \frac{1}{n}\sum_{i=1}^{n} K_{G(c_\ell,b)}\,(X_i) \overset{p}{\to} \frac{f_+\,(c_\ell) + f_-\,(c_\ell)}{2},$$

we can obtain another estimator $\widetilde{\mathbf{V}}_2\,(\mathbf{c}) := \mathrm{diag}\left\{\tilde{V}_2\,(c_1),\ldots,\tilde{V}_2\,(c_L)\right\}$, where

$$\tilde{V}_2\,(c_\ell) := \upsilon\,(\delta)\left\{\frac{2\hat{f}_G\,(c_\ell)}{\sqrt{\pi}c_\ell^{1/2}}\right\}.$$

### 5.3.3    Estimation of the Entire Density in the Presence of Multiple Discontinuity Points

#### 5.3.3.1    Density Estimation by Truncated Kernels

Graphical analyses including inspections of densities of running variables are strongly encouraged in empirical studies on RDD. Below we consider the problem of estimating $f$ for a given design point $x$ that differs from $L$ prespecified cutoffs $c_1,\ldots,c_L$. It turns out that $f\,(x)$ for $x \in [0, c_1)$ and for $x \in (c_\ell, c_{\ell+1})$, $\ell = 1,\ldots,L$

can be consistently estimated by

$$\hat{f}_{[0,c_1)}(x) := \frac{1}{n}\sum_{i=1}^{n} K_{G(x,b;0,c_1)}(X_i) \text{ and } \hat{f}_{(c_\ell,c_{\ell+1})}(x) := \frac{1}{n}\sum_{i=1}^{n} K_{G(x,b;c_\ell,c_{\ell+1})}(X_i),$$

respectively, regardless of whether $f$ may be continuous or discontinuous at the cutoffs.

The density estimators should be consistent at design points other than $L$ cutoffs. Therefore, after strengthening Assumption 5.8 suitably, we document a theorem on consistency of the density estimators.

**Assumption 5.10** The second-order derivative of $f$ is continuous and bounded on $\mathbb{R}_+ \setminus \{c_1, \ldots, c_L\}$.

**Theorem 5.5** *Suppose that Assumptions 5.7, 5.10, and 5.9 hold. Then, for $\ell = 1, \ldots, L$, as $n \to \infty$,*

$$Bias\left\{\hat{f}_{(c_\ell,c_{\ell+1})}(x)\right\} = \left\{f^{(1)}(x) + \frac{x}{2}f^{(2)}(x)\right\}b + o(b), \text{ and}$$

$$Var\left\{\hat{f}_{(c_\ell,c_{\ell+1})}(x)\right\} = \frac{1}{nb^{1/2}}\frac{f(x)}{2\sqrt{\pi}x^{1/2}} + o\left(n^{-1}b^{-1/2}\right).$$

*On the other hand, for $\ell = 0$, as $n \to \infty$,*

$$Bias\left\{\hat{f}_{[0,c_1)}(x)\right\} = \left\{f^{(1)}(x) + \frac{x}{2}f^{(2)}\right\}b + o(b), \text{ and}$$

$$Var\left\{\hat{f}_{[0,c_1)}(x)\right\} = \begin{cases} \frac{1}{nb^{1/2}}\frac{f(x)}{2\sqrt{\pi}x^{1/2}} + o\left(n^{-1}b^{-1/2}\right) & \text{if } x/b \to \infty \\ \frac{1}{nb}\frac{\Gamma(2\kappa+1)}{2^{2\kappa+1}\Gamma^2(\kappa+1)}f(x) + o\left(n^{-1}b^{-1}\right) & \text{if } x/b \to \kappa \in (0,\infty) \end{cases}.$$

*Proof* See Sect. 5.5.2.

Theorem 5.5 indicates no adversity when $f(x)$ for $x \in (c_\ell, c_{\ell+1})$ is estimated by $\hat{f}_{(c_\ell,c_{\ell+1})}(x)$; although only the bias-variance trade-off is provided, asymptotic normality of the estimators can be established similarly to Theorem 5.4. Observe that $\hat{f}_{(c_\ell,c_{\ell+1})}(x)$ admit the same bias and variance expansions as $\hat{f}_G(x)$ does. A rationale is that as the design point $x$ moves away from both of two adjacent cutoffs $c_\ell$ and $c_{\ell+1}$, data points tend to lie on both sides of $x$ and each truncated kernel is likely to behave like the G kernel.

### 5.3.3.2 Properties of $\hat{f}_{[0,c_1)}(x)$ at the Origin

Properties of density estimators at the origin that are discussed in Sect. 2.5 also apply to $\hat{f}_{[0,c_1)}(x)$. First, if $f(0) < \infty$, then under Assumptions 5.7, 5.10, and 5.9 the bias and variance of $\hat{f}_{[0,c_1)}(0)$ can be approximated by

$$Bias\left\{\hat{f}_{[0,c_1)}(0)\right\} = f^{(1)}(0)b + o(b), \text{ and}$$

$$Var\left\{\hat{f}_{[0,c_1)}(0)\right\} = \begin{cases} \frac{1}{nb}\frac{f(0)}{2} + o\left(n^{-1}b^{-1}\right) & \text{if } f(0) > 0 \\ \frac{1}{n}\frac{f^{(1)}(0)}{4} + o\left(n^{-1}\right) & \text{if } f(0) = 0 \text{ and } f^{(1)}(0) > 0 \end{cases},$$

respectively. Observe that these results are the same as those for $\hat{f}_G(x)$.

Second, clusterings of observations near the boundary are observed even in the study of RDD (e.g., Fig. 5 of McCrary 2008). The following two theorems document weak consistency and the relative convergence of $\hat{f}_{[0,c_1)}(x)$ when $f(x)$ is unbounded at $x = 0$.

**Theorem 5.6** (Funke and Hirukawa 2017a, Theorem 3)

*If $f(x)$ is unbounded at $x = 0$, Assumption 5.7 holds, and $b + \left(nb^2\right)^{-1} \to 0$ as $n \to \infty$, then $\hat{f}_{[0,c_1)}(0) \xrightarrow{p} \infty$.*

**Theorem 5.7** (Funke and Hirukawa 2017a, Theorem 4)

*Suppose that $f(x)$ is unbounded at $x = 0$ and continuously differentiable in the neighborhood of the origin. In addition, if Assumption 5.7 holds and $b + \left\{nb^2 f(x)\right\}^{-1} \to 0$ as $n \to \infty$ and $x \to 0$, then*

$$\left|\frac{\hat{f}_{[0,c_1)}(x) - f(x)}{f(x)}\right| \xrightarrow{p} 0$$

*as $x \to 0$.*

As seen in Sect. 2.5, the weak consistency and relative convergence for densities unbounded at the origin are peculiar to the density estimators smoothed by the B, G, and GG kernels. The theorems ensure that $\hat{f}_{[0,c_1)}(x)$ is also a proper estimate for unbounded densities. We can deduce from Theorems 5.5–5.7 that all in all, appealing properties of $\hat{f}_G$ are inherited to $\hat{f}_{[0,c_1)}$ and $\hat{f}_{(c_\ell,c_{\ell+1})}$, $\ell = 1, \ldots, L$.

## 5.4   Smoothing Parameter Selection

We conclude this chapter by discussing the problem of choosing the smoothing parameter $b$ for asymmetric kernel-based testing. Little is known about the problem, despite its importance. Accordingly, some authors simply adopt the choice method implied by density estimation. Fernandes and Grammig (2005) employ a method similar to Silverman's (1986) rule of thumb for the test of a parametric form in ACD models, whereas Fernandes et al. (2015) adjust the value chosen via CV for the SSST.

These approaches cannot be justified in theory or practice, because the value of $b$ chosen in an estimation–optimal criterion may not be equally optimal for inference. In light of test optimality, Hirukawa and Sakudo (2016) and Funke and Hirukawa (2017a) tailor the approach by Kulasekera and Wang (1998) to the SSST and the test of discontinuity in densities, respectively. Below we briefly discuss the former. In Chap. 6, the latter will be extended to the case of the joint test of discontinuity in densities at multiple cutoffs.

The main idea in Kulasekera and Wang (1998) is grounded on subsampling. Without loss of generality, it is assumed that $\{X_i\}_{i=1}^{n_1}$ and $\{Y_i\}_{i=1}^{n_2}$ are ordered samples, where $n_1 \neq n_2$ may be the case. Then, the entire sample $\{\{X_i\}_{i=1}^{n_1}, \{Y_i\}_{i=1}^{n_2}\}$ can be split into $M$ subsamples, where $M = M_{n_1}$ is a nonstochastic sequence that satisfies $1/M + M/n_1 \to 0$ as $n_1 \to \infty$. Given such $M$ and $(k_1, k_2) := (\lfloor n_1/M \rfloor, \lfloor n_2/M \rfloor)$, the $m$th subsample is defined as $\{\{X_{m+(i-1)M}\}_{i=1}^{k_1}, \{Y_{m+(i-1)M}\}_{i=1}^{k_2}\}$, $m = 1, \ldots, M$. The test statistic using the $m$th subsample becomes

$$T_{k_1,k_2,j}(m) := \frac{k_1 b^{1/4} I_{k_1,k_2}(m)}{\hat{\sigma}_\lambda(m)}, \quad m = 1, \ldots, M,$$

where $I_{k_1,k_2,j}(m)$ and $\hat{\sigma}_\lambda^2(m)$ are the subsample analogs to (5.1) and (5.2), respectively. Also, denote the set of admissible values for $b = b_{n_1}$ as $H_{n_1} := \left[\underline{B} n_1^{-q}, \overline{B} n_1^{-q}\right]$ for some prespecified exponent $q \in (2/5, 1)$ and two constants $0 < \underline{B} < \overline{B} < \infty$. Moreover, let

$$\hat{\pi}_M(b_{k_1}) := \frac{1}{M} \sum_{m=1}^M \mathbf{1}\left\{T_{k_1,k_2}(m) > c_m(\alpha)\right\},$$

where $c_m(\alpha)$ is the critical value for the size $\alpha$ test using the $m$th subsample. We pick the power-maximizing value

$$\hat{b}_{k_1} = \hat{B} k_1^{-q} = \arg\max_{b_{k_1} \in H_{k_1}} \hat{\pi}_M(b_{k_1}),$$

and the smoothing parameter value $\hat{b}_{n_1} := \hat{B} n_1^{-q}$ follows.

In practice, the test-optimal $\hat{b}_{n_1}$ may be chosen in the following five steps. Step 1 reflects that $M$ should be divergent but smaller than both $n_1$ and $n_2$ in finite samples. Step 3 follows from the implementation methods in Kulasekera and Wang (1998). Finally, Step 4 considers that there may be more than one maximizer of $\hat{\pi}_M(b_{k_1})$.

**Step 1**: Pick $M := \min\left\{\lfloor n_1^\delta \rfloor, \lfloor n_2^\delta \rfloor\right\}$ for some $\delta \in (0, 1)$.

**Step 2**: Make $M$ subsamples of sizes $(k_1, k_2) := (\lfloor n_1/M \rfloor, \lfloor n_2/M \rfloor)$.

**Step 3**: Choose two constants $0 < \underline{H} < \overline{H} < 1$ and define $H_{k_1} := [\underline{H}, \overline{H}]$.

**Step 4**: Set $c_m(\alpha) \equiv z_\alpha$ (i.e., $\Pr\{N(0, 1) > z_\alpha\} = \alpha$) and find $\hat{b}_{k_1} = \inf\left\{\arg\max_{b_{k_1} \in H_{k_1}} \hat{\pi}_M(b_{k_1})\right\}$ by a grid search.

**Step 5**: Recover $\hat{B} = \hat{b}_{k_1} k_1^q$ for some $q \in (2/5, 1)$ and obtain $\hat{b}_{n_1} = \hat{B} n_1^{-q}$.

## 5.5  Technical Proofs

### 5.5.1  Proof of Theorem 5.4

The proof requires the following lemmata.

**Lemma 5.1** *Let $\beta_1$, $\beta_2$, and $\beta_3$ be sequences such that $0 < \beta_1 < \beta_2 < \beta_3$ and $\beta_1, \beta_2, \beta_3 \to \infty$. Then,*

$$\frac{\gamma(\beta_2+1,\beta_3)}{\Gamma(\beta_2+1)} = 1 + O\left[\beta_2^{1/2} \exp\left\{\beta_2 \log\left(\frac{e}{\eta e^{1/\eta}}\right)\right\}\right]$$

$$= 1 + o(1),$$

$$\frac{\gamma(\beta_2+2,\beta_3)}{\Gamma(\beta_2+1)} = \beta_2 + 1 + O\left[\beta_2^{3/2} \exp\left\{\beta_2 \log\left(\frac{e}{\eta e^{1/\eta}}\right)\right\}\right]$$

$$= \beta_2 + 1 + o(1),$$

$$\frac{\gamma(\beta_2+3,\beta_3)}{\Gamma(\beta_2+1)} = \beta_2^2 + 3\beta_2 + 2 + O\left[\beta_2^{5/2} \exp\left\{\beta_2 \log\left(\frac{e}{\eta e^{1/\eta}}\right)\right\}\right]$$

$$= \beta_2^2 + 3\beta_2 + 2 + o(1),$$

$$\frac{\gamma(\beta_2+1,\beta_1)}{\Gamma(\beta_2+1)} = O\left[\beta_2^{1/2} \exp\left\{\beta_2 \log\left(\frac{\eta' e}{e^{\eta'}}\right)\right\}\right] = o(1),$$

$$\frac{\gamma(\beta_2+2,\beta_1)}{\Gamma(\beta_2+1)} = O\left[\beta_2^{3/2} \exp\left\{\beta_2 \log\left(\frac{\eta' e}{e^{\eta'}}\right)\right\}\right] = o(1), \text{ and}$$

$$\frac{\gamma(\beta_2+3,\beta_1)}{\Gamma(\beta_2+1)} = O\left[\beta_2^{5/2} \exp\left\{\beta_2 \log\left(\frac{\eta' e}{e^{\eta'}}\right)\right\}\right] = o(1),$$

*where $\eta := \beta_2/\beta_3 \in (0,1)$ and $\eta' := \beta_1/\beta_2 \in (0,1)$ so that $e/\left(\eta e^{1/\eta}\right), \eta' e/e^{\eta'} \in (0,1)$.*

**Lemma 5.2** *Under Assumptions 5.7–5.9, as $n \to \infty$,*

$$Bias\left\{\hat{f}_\pm(c_\ell)\right\}$$

$$= \pm\sqrt{\frac{2}{\pi}} c_\ell^{1/2} f_\pm^{(1)}(c_\ell) b^{1/2} + \left\{\left(1 - \frac{4}{3\pi}\right) f_\pm^{(1)}(c_\ell) + \frac{c_\ell}{2} f_\pm^{(2)}(c_\ell)\right\} b + o(b), \text{ and}$$

$$Var\left\{\hat{f}_\pm(c_\ell)\right\}$$

$$= \frac{1}{nb^{1/2}} \frac{f_\pm(c_\ell)}{\sqrt{\pi} c_\ell^{1/2}} + o\left(n^{-1}b^{-1/2}\right)$$

*for $\ell = 1, \ldots, L$.*

**Lemma 5.3**  *Under Assumptions 5.7–5.9, as $n \to \infty$,*

$$Cov\left\{\tilde{f}_+(c_\ell), \tilde{f}_-(c_{\ell+m})\right\} = \begin{cases} O\left(n^{-1}\right) & \text{if } m = 1 \\ 0 & \text{if } m \neq 1 \end{cases}$$

*for $\ell = 1, \ldots, L$ and $m = 0, 1, \ldots, L - \ell$.*

**Proof of Lemma 5.1**

The first equality for each statement immediately establishes the second one because of an exponentially decaying rate in the big $O$ term. Then, we may concentrate on demonstrating the first equalities. These can be obtained by using (2.16), (2.17) and arguments in the Proof of Theorem 2 in Funke and Hirukawa (2017b). ∎

**Proof of Lemma 5.2**

Proposition 1 of Funke and Hirukawa (2017a) has already established the bias-variance trade-off on $\hat{f}_-(c_1)$ and $\hat{f}_+(c_L)$. To save space, without loss of generality, we focus only on approximating $E\left\{\hat{f}_+(c_1)\right\}$ and $Var\left\{\hat{f}_+(c_1)\right\}$. Bias and variance approximations to other estimators can be obtained in a similar manner.

**Bias.** By the change of variable $v := u/b$,

$$E\left\{\hat{f}_+(c_1)\right\} = \int_{c_1}^{c_2} \frac{u^{c_1/b} \exp(-u/b) f(u)}{b^{c_1/b+1} \left\{\gamma(c_1/b + 1, c_2/b) - \gamma(c_1/b + 1, c_1/b)\right\}} du$$

$$= \int_{a_1}^{a_2} f(bv) \left\{\frac{v^{a_1} \exp(-v)}{\gamma(a_1 + 1, a_2) - \gamma(a_1 + 1, a_1)}\right\} dv,$$

where $(a_1, a_2) := (c_1/b, c_2/b)$, and the object inside brackets of the right-hand side is a pdf on the interval $[a_1, a_2)$. Then, a second-order Taylor expansion of $f(bv)$ around $bv = c_1$ (from above) yields

$$E\left\{\hat{f}_+(c_1)\right\}$$
$$= f_+(c_1) + bf_+^{(1)}(c_1) \left\{\frac{\gamma(a_1 + 2, a_2) - \gamma(a_1 + 2, a_1)}{\gamma(a_1 + 1, a_2) - \gamma(a_1 + 1, a_1)} - a_1\right\}$$
$$+ \frac{b^2}{2} f_+^{(2)}(c_1) \left\{\frac{\gamma(a_1 + 3, a_2) - \gamma(a_1 + 3, a_1)}{\gamma(a_1 + 1, a_2) - \gamma(a_1 + 1, a_1)}\right.$$
$$\left. - 2a_1 \frac{\gamma(a_1 + 2, a_2) - \gamma(a_1 + 2, a_1)}{\gamma(a_1 + 1, a_2) - \gamma(a_1 + 1, a_1)} + a_1^2\right\} + R_{\hat{f}_+(c_1)},$$

where

$$R_{\hat{f}_+(c_1)} := \frac{b^2}{2} \int_{a_1}^{a_2} \left\{f^{(2)}(\xi) - f^{(2)}(c_1)\right\} (v - a_1)^2 \left\{\frac{v^{a_1} \exp(-v)}{\gamma(a_1 + 1, a_2) - \gamma(a_1 + 1, a_1)}\right\} dv$$

is the remainder term with $\xi = \alpha\,(bv) + (1 - \alpha)\,c_1$ for some $\alpha \in (0, 1)$.

Observe that $a_1 < a_2$ and $a_1, a_2 \to \infty$ as $n \to \infty$. Putting $(\beta_2, \beta_3) = (a_1, a_2)$ in Lemma 5.1, using (2.17) and equation (A6) of Funke and Hirukawa (2017b) repeatedly, and making some straightforward but tedious calculations, we have

$$
\frac{\gamma\,(a_1 + 2, a_2) - \gamma\,(a_1 + 2, a_1)}{\gamma\,(a_1 + 1, a_2) - \gamma\,(a_1 + 1, a_1)} - a_1
$$

$$
= \sqrt{\frac{2}{\pi}}\,a_1^{1/2} + \left(1 - \frac{4}{3\pi}\right) + O\left(a_1^{-1/2}\right)
$$

$$
= \sqrt{\frac{2}{\pi}}\left(\frac{c_1}{b}\right)^{1/2} + \left(1 - \frac{4}{3\pi}\right) + O\left(b^{1/2}\right),\ \text{and}
$$

$$
\frac{\gamma\,(a_1 + 3, a_2) - \gamma\,(a_1 + 3, a_1)}{\gamma\,(a_1 + 1, a_2) - \gamma\,(a_1 + 1, a_1)} - 2a_1 \frac{\gamma\,(a_1 + 2, a_2) - \gamma\,(a_1 + 2, a_1)}{\gamma\,(a_1 + 1, a_2) - \gamma\,(a_1 + 1, a_1)} + a_1^2
$$

$$
= a_1 + O\left(a_1^{1/2}\right)
$$

$$
= \frac{c_1}{b} + O\left(b^{-1/2}\right).
$$

A similar argument to the Proof of Proposition 1 in Funke and Hirukawa (2017b) can also establish that $R_{\hat{f}_+(c_1)} = o\,(b)$. Then, the bias approximation immediately follows.

**Variance.** By

$$
Var\left\{\hat{f}_+\,(c_1)\right\} = \frac{1}{n} E\left\{K^2_{G(c_1,b;c_1,c_2)}\,(X_i)\right\} + O\left(n^{-1}\right),
$$

we may concentrate on approximating $E\left\{K^2_{G(c_1,b;c_1,c_2)}\,(X_i)\right\}$. A straightforward calculation and the change of variable $w := 2u/b$ yield

$$
E\left\{K^2_{G(c_1,b;c_1,c_2)}\,(X_i)\right\}
$$

$$
= \int_{c_1}^{c_2} \frac{u^{2c_1/b}\exp\,(-2u/b)\,f\,(u)}{b^{2(c_1/b+1)}\,\{\gamma\,(c_1/b + 1, c_2/b) - \gamma\,(c_1/b + 1, c_2/b)\}^2}\,du
$$

$$
= \left\{\frac{b^{-1}\Gamma\,(2a_1 + 1)}{2^{2a_1+1}\Gamma^2\,(a_1 + 1)}\right\}\left\{\frac{\gamma\,(2a_1 + 1, 2a_2)}{\Gamma\,(2a_1 + 1)} - \frac{\gamma\,(2a_1 + 1, 2a_1)}{\Gamma\,(2a_1 + 1)}\right\}
$$

$$
\times \left\{\frac{\gamma\,(a_1 + 1, a_2)}{\Gamma\,(a_1 + 1)} - \frac{\gamma\,(a_1 + 1, a_1)}{\Gamma\,(a_1 + 1)}\right\}^{-2}
$$

$$
\times \int_{2a_1}^{2a_2} f\left(\frac{bw}{2}\right)\left\{\frac{w^{2a_1}\exp\,(-w)}{\gamma\,(2a_1 + 1, 2a_2) - \gamma\,(2a_1 + 1, 2a_1)}\right\}\,dw,
$$

where $(a_1, a_2) = (c_1/b, c_2/b)$ as in the bias approximation part, and the object inside brackets in the integral on the right-hand side is again a pdf on the interval $[2a_1, 2a_2)$.

As before, the integral part can be approximated by $f_+ (c_1) + O\left(b^{1/2}\right)$. Applying Lemma 5.1, equation (A6) of Funke and Hirukawa (2017b), and the argument on p.474 of Chen (2000) and recognizing that $a_1 = c_1/b$ also yield

$$
\left\{ \frac{b^{-1}\Gamma\left(2a_1 + 1\right)}{2^{2a_1+1}\Gamma^2\left(a_1 + 1\right)} \right\} \left\{ \frac{\gamma\left(2a_1 + 1, 2a_2\right)}{\Gamma\left(2a_1 + 1\right)} - \frac{\gamma\left(2a_1 + 1, 2a_1\right)}{\Gamma\left(2a_1 + 1\right)} \right\}
$$
$$
\times \left\{ \frac{\gamma\left(a_1 + 1, a_2\right)}{\Gamma\left(a_1 + 1\right)} - \frac{\gamma\left(a_1 + 1, a_1\right)}{\Gamma\left(a_1 + 1\right)} \right\}^{-2}
$$
$$
= \frac{b^{-1/2}}{\sqrt{\pi}c_1^{1/2}} + o\left(b^{-1/2}\right).
$$

Therefore, the variance approximation is also demonstrated, which completes the proof. ∎

**Proof of Lemma 5.3**

It is the case that $Cov\left\{\tilde{f}_+ (c_\ell), \tilde{f}_- (c_{\ell+m})\right\} = 0$ if $m = 0$ or $m \geq 2$, because the subsamples used for $\tilde{f}_+ (c_\ell)$ and $\tilde{f}_- (c_{\ell+m})$ do not overlap. Our remaining task is to demonstrate, without loss of generality, that $Cov\left\{\tilde{f}_+ (c_1), \tilde{f}_- (c_2)\right\} = O\left(n^{-1}\right)$.

Let $I_{\pm,b} (c_\ell) := E\left\{\hat{f}_{\pm,b} (c_\ell)\right\}$, $Z_{\pm,b} (c_\ell) := \hat{f}_{\pm,b} (c_\ell) - E\left\{\hat{f}_{\pm,b} (c_\ell)\right\}$, and $W_\pm (c_\ell) := Z_{\pm,b} (c_\ell) - \delta^{1/2} Z_{\pm,b/\delta} (c_\ell)$. Then, by a similar argument to the Proof of Theorem 1 in Hirukawa (2010),

$$
\tilde{f}_\pm (c_\ell) = \left\{I_{\pm,b} (c_\ell)\right\}^{1/\left(1-\delta^{1/2}\right)} \left\{I_{\pm,b/\delta} (c_\ell)\right\}^{-\delta^{1/2}/\left(1-\delta^{1/2}\right)}
$$
$$
+ \left(\frac{1}{1 - \delta^{1/2}}\right) W_\pm (c_\ell) + R_{\tilde{f}_\pm(c_\ell)},
$$

where the remainder term $R_{\tilde{f}_\pm(c_\ell)}$ is of smaller order than $Z_{\pm,b} (c_\ell)$ and $Z_{\pm,b/\delta} (c_\ell)$. It follows that the proof is completed if all the followings are proven to be true:

$$
E\left\{Z_{+,b} (c_1) Z_{-,b} (c_2)\right\} = O\left(n^{-1}\right); \tag{5.3}
$$
$$
E\left\{Z_{+,b} (c_1) Z_{-,b/\delta} (c_2)\right\} = O\left(n^{-1}\right); \tag{5.4}
$$
$$
E\left\{Z_{+,b/\delta} (c_1) Z_{-,b} (c_2)\right\} = O\left(n^{-1}\right); \text{ and} \tag{5.5}
$$
$$
E\left\{Z_{+,b/\delta} (c_1) Z_{-,b/\delta} (c_2)\right\} = O\left(n^{-1}\right). \tag{5.6}
$$

Observe that (5.6) automatically holds once (5.3) is shown. In addition, the proof strategies for (5.4) and (5.5) are basically the same as that of (5.3). Therefore, we demonstrate only (5.3) below.

By the definitions of $Z_{+,b}(c_1)$ and $Z_{-,b}(c_2)$,

$$E\left\{Z_{+,b}(c_1)\,Z_{-,b}(c_2)\right\}$$
$$= \frac{1}{n}E\left\{K_{G(c_1,b;c_1,c_2)}(X_i)\,K_{G(c_2,b;c_1,c_2)}(X_i)\right\} + O\left(n^{-1}\right).$$

A straightforward calculation and the change of variable $w := 2u/b$ yield

$$E\left\{K_{G(c_1,b;c_1,c_2)}(X_i)\,K_{G(c_2,b;c_1,c_2)}(X_i)\right\}$$
$$= \int_{c_1}^{c_2} \frac{u^{c_1/b+c_2/b}\exp(-2u/b)\,f(u)}{b^{c_1/b+c_2/b+2}\left\{\gamma\left(\frac{c_1}{b}+1,\frac{c_2}{b}\right)-\gamma\left(\frac{c_1}{b}+1,\frac{c_1}{b}\right)\right\}\left\{\gamma\left(\frac{c_2}{b}+1,\frac{c_2}{b}\right)-\gamma\left(\frac{c_2}{b}+1,\frac{c_1}{b}\right)\right\}}\,du$$
$$= \left\{\frac{b^{-1}\Gamma(a_1+a_2+1)}{2^{a_1+a_2+1}\Gamma(a_1+1)\Gamma(a_2+1)}\right\}\left\{\frac{\gamma(a_1+a_2+1,2a_2)}{\Gamma(a_1+a_2+1)}-\frac{\gamma(a_1+a_2+1,2a_1)}{\Gamma(a_1+a_2+1)}\right\}$$
$$\times\left\{\frac{\gamma(a_1+1,a_2)}{\Gamma(a_1+1)}-\frac{\gamma(a_1+1,a_1)}{\Gamma(a_1+1)}\right\}^{-1}\left\{\frac{\gamma(a_2+1,a_2)}{\Gamma(a_2+1)}-\frac{\gamma(a_2+1,a_1)}{\Gamma(a_2+1)}\right\}^{-1}$$
$$\times\int_{2a_1}^{2a_2} f\left(\frac{bw}{2}\right)\left\{\frac{w^{a_1+a_2}\exp(-w)}{\gamma(a_1+a_2+1,2a_2)-\gamma(a_1+a_2+1,2a_1)}\right\}dw,$$

where $(a_1,a_2)=(c_1/b,c_2/b)$ as before. Obviously, the integral part is at most $O(1)$. Lemma 5.1 and equation (A6) of Funke and Hirukawa (2017b) also give

$$\frac{\gamma(a_1+1,a_2)}{\Gamma(a_1+1)}-\frac{\gamma(a_1+1,a_1)}{\Gamma(a_1+1)}=\frac{1}{2}+o(1)=O(1),\text{ and}$$
$$\frac{\gamma(a_2+1,a_2)}{\Gamma(a_2+1)}-\frac{\gamma(a_2+1,a_1)}{\Gamma(a_2+1)}=\frac{1}{2}+o(1)=O(1).$$

Because $2a_1 < a_1+a_2 < 2a_2$ and $a_1,a_2\to\infty$ as $n\to\infty$,

$$\frac{\gamma(a_1+a_2+1,2a_2)}{\Gamma(a_1+a_2+1)}-\frac{\gamma(a_1+a_2+1,2a_1)}{\Gamma(a_1+a_2+1)}=1+o(1)=O(1).$$

Finally, (2.16) yields

$$\frac{b^{-1}\Gamma(a_1+a_2+1)}{2^{a_1+a_2+1}\Gamma(a_1+1)\Gamma(a_2+1)}$$
$$=\frac{b^{-1}}{2\sqrt{2\pi}}\left(\frac{1}{a_1}+\frac{1}{a_2}\right)^{1/2}\left(\frac{a_1+a_2}{2a_1}\right)^{a_1}\left(\frac{a_1+a_2}{2a_2}\right)^{a_2}\{1+o(1)\},$$

It is easy to see that $(1/a_1+1/a_2)^{1/2}=O\left(b^{1/2}\right)$. Denoting $\rho=a_1/a_2\in(0,1)$ as before, we also have

$$\left(\frac{a_1+a_2}{2a_1}\right)^{a_1}\left(\frac{a_1+a_2}{2a_2}\right)^{a_2}=\exp\left[a_2\log\left\{\left(\frac{1+\rho}{2\rho}\right)^{\rho}\left(\frac{1+\rho}{2}\right)\right\}\right],$$

where $\{(1 + \rho) / (2\rho)\}^\rho \{(1 + \rho) / 2\} \in (1/2, 1)$ so that the right-hand side converges to zero at an exponential rate. Therefore,

$$\frac{b^{-1}\Gamma(a_1 + a_2 + 1)}{2^{a_1 + a_2 + 1}\Gamma(a_1 + 1)\Gamma(a_2 + 1)} = o(1),$$

and thus $E\{K_{G(c_1, b; c_1, c_2)}(X_i) K_{G(c_2, b; c_1, c_2)}(X_i)\} = o(1)$, which establishes (5.3). ∎

**Proof of Theorem 5.4**

**(i)** This part is obvious by Lemmata 5.1 and 5.2 and the bias and variance approximation techniques for the TS-MBC estimation (e.g., the Proof of Theorem 1 in Hirukawa 2010).

**(ii)** Let $\mathbf{W}_\pm(\mathbf{c}) := (W_\pm(c_1), \ldots, W_\pm(c_L))^\top$. Then, for an arbitrary vector $\mathbf{t} := (t_1, \ldots, t_L)^\top \in \mathbb{R}^L$,

$$\sqrt{nb^{1/2}}\{\mathbf{t}^\top \widetilde{\mathbf{J}}(\mathbf{c}) - \mathbf{t}^\top \mathbf{J}(\mathbf{c})\}$$
$$= \sqrt{nb^{1/2}}\left[\mathbf{t}^\top \widetilde{\mathbf{J}}(\mathbf{c}) - E\{\mathbf{t}^\top \widetilde{\mathbf{J}}(\mathbf{c})\}\right] + \sqrt{nb^{1/2}}\left[E\{\mathbf{t}^\top \widetilde{\mathbf{J}}(\mathbf{c})\} - \mathbf{t}^\top \mathbf{J}(\mathbf{c})\right]$$
$$= \sqrt{nb^{1/2}}\left(\frac{1}{1 - \delta^{1/2}}\right)\mathbf{t}^\top\{\mathbf{W}_+(\mathbf{c}) - \mathbf{W}_-(\mathbf{c})\}$$
$$+ \sqrt{nb^{1/2}}\mathbf{t}^\top\{\mathbf{B}(\mathbf{c})b + o(b)\} + o_p(1),$$

where $\sqrt{nb^{1/2}}\mathbf{t}^\top\{\mathbf{B}(\mathbf{c})b + o(b)\} = o(1)$ if $nb^{5/2} \to 0$. It follows from the Cramér-Wold device that the proof of this part is completed if

$$\sqrt{nb^{1/2}}\left(\frac{1}{1 - \delta^{1/2}}\right)\mathbf{t}^\top\{\mathbf{W}_+(\mathbf{c}) - \mathbf{W}_-(\mathbf{c})\} \overset{d}{\to} N\left(0, \mathbf{t}^\top \mathbf{V}(\mathbf{c})\mathbf{t}\right).$$

Because Lemmata 5.2 and 5.3 have already established the asymptotic variance of the left-hand side, we only need to demonstrate Liapunov's condition for this term.
    Let

$$\mathbf{W}_\pm(\mathbf{c}) := \sum_{i=1}^n \mathbf{W}_{\pm, i}(\mathbf{c}) := \sum_{i=1}^n \{\mathbf{Z}_{\pm, b, i}(\mathbf{c}) - \delta^{1/2}\mathbf{Z}_{\pm, b/\delta, i}(\mathbf{c})\},$$

where

$$\mathbf{Z}_{+,b,i}\left(\mathbf{c}\right)=\frac{1}{n}\begin{bmatrix} K_{G(c_1,b;c_1,c_2)}\left(X_i\right)-E\left\{K_{G(c_1,b;c_1,c_2)}\left(X_i\right)\right\} \\ \vdots \\ K_{G(c_L,b;c_L,\infty)}\left(X_i\right)-E\left\{K_{G(c_L,b;c_L,\infty)}\left(X_i\right)\right\} \end{bmatrix} \quad\text{and}$$

$$\mathbf{Z}_{-,i,b}\left(\mathbf{c}\right)=\frac{1}{n}\begin{bmatrix} K_{G(c_1,b;0,c_1)}\left(X_i\right)-E\left\{K_{G(c_1,b;0,c_1)}\left(X_i\right)\right\} \\ \vdots \\ K_{G(c_L,b;c_{L-1},c_L)}\left(X_i\right)-E\left\{K_{G(c_L,b;c_{L-1},c_L)}\left(X_i\right)\right\} \end{bmatrix}.$$

Then,

$$\sqrt{nb^{1/2}}\left(\frac{1}{1-\delta^{1/2}}\right)\mathbf{t}^{\top}\left\{\mathbf{W}_+\left(\mathbf{c}\right)-\mathbf{W}_-\left(\mathbf{c}\right)\right\}$$

$$=\sum_{i=1}^{n}\sqrt{\frac{b^{1/2}}{n}}\left(\frac{1}{1-\delta^{1/2}}\right)\mathbf{t}^{\top}\left\{\mathbf{W}_{+,i}\left(\mathbf{c}\right)-\mathbf{W}_{-,i}\left(\mathbf{c}\right)\right\}.$$

Further, denote $Y_i:=\sqrt{b^{1/2}/n}\left(1-\delta^{1/2}\right)^{-1}\mathbf{t}^{\top}\left\{\mathbf{W}_{+,i}\left(\mathbf{c}\right)-\mathbf{W}_{-,i}\left(\mathbf{c}\right)\right\}$. Then, $E\left|Y_i\right|^3$ is bounded by

$$\frac{b^{3/4}}{n^{3/2}}\frac{\|\mathbf{t}\|^3}{\left(1-\delta^{1/2}\right)^3}$$

$$\times E\left\{\left(\left\|\mathbf{Z}_{+,i,b}\left(\mathbf{c}\right)\right\|+\left\|\mathbf{Z}_{+,i,b/\delta}\left(\mathbf{c}\right)\right\|+\left\|\mathbf{Z}_{-,i,b}\left(\mathbf{c}\right)\right\|+\left\|\mathbf{Z}_{-,i,b/\delta}\left(\mathbf{c}\right)\right\|\right)^3\right\}.$$

Similarly to Lemma A1 of Funke and Hirukawa (2017b), the expected value part is shown to be at most $O\left(b^{-1}\right)$. Hence, $E\left|Y_i\right|^3=O\left(n^{-3/2}b^{-1/4}\right)$. It is also straightforward to see that $Var\left(Y_i\right)=O\left(n^{-1}\right)$. Therefore,

$$\frac{\sum_{i=1}^{n}E\left|Y_i\right|^3}{\left\{\sum_{i=1}^{n}Var\left(Y_i\right)\right\}^{3/2}}=O\left(n^{-1/2}b^{-1/4}\right)\to 0,$$

or Liapunov's condition holds. This completes the proof.  ∎

### 5.5.2  Proof of Theorem 5.5

Because the results on $\hat{f}_{[0,c_1)}\left(x\right)$ and $\hat{f}_{(c_L,\infty)}\left(x\right)$ have been made available as Theorem 2 of Funke and Hirukawa (2017a), it suffices to establish bias and variance approximations to $\hat{f}_{(c_\ell,c_{\ell+1})}\left(x\right)$, $\ell=1,\ldots,L-1$. Without loss of generality, we put $\ell=1$.

**Bias.** By the change of variable $v:=u/b$,

$$E\left\{\hat{f}_{(c_1,c_2)}(x)\right\} = \int_{c_1}^{c_2} \frac{u^{x/b}\exp(-u/b)\,f(u)}{b^{x/b+1}\left\{\gamma(x/b+1,c_2/b)-\gamma(x/b+1,c_1/b)\right\}}du$$

$$= \int_{a_1}^{a_2} f(bv)\left\{\frac{v^z\exp(-v)}{\gamma(z+1,a_2)-\gamma(z+1,a_1)}\right\}dv,$$

where $(a_1,a_2):=(c_1/b,c_2/b)$, $z:=x/b\in(a_1,a_2)$, and the object inside brackets of the right-hand side is a pdf on the interval $(a_1,a_2)$. Then, a second-order Taylor expansion of $f(bv)$ around $bv=x$ yields

$$E\left\{\hat{f}_{(c_1,c_2)}(x)\right\}$$
$$=f(x)+bf^{(1)}(x)\left\{\frac{\gamma(z+2,a_2)-\gamma(z+2,a_1)}{\gamma(z+1,a_2)-\gamma(z+1,a_1)}-z\right\}$$
$$+\frac{b^2}{2}f^{(2)}(x)\left\{\frac{\gamma(z+3,a_2)-\gamma(z+3,a_1)}{\gamma(z+1,a_2)-\gamma(z+1,a_1)}\right.$$
$$\left.-2z\frac{\gamma(z+2,a_2)-\gamma(z+2,a_1)}{\gamma(z+1,a_2)-\gamma(z+1,a_1)}+z^2\right\}+R_{\hat{f}_{(c_1,c_2)}(x)},$$

where

$$R_{\hat{f}_{(c_1,c_2)}(x)}:=\frac{b^2}{2}\int_{a_1}^{a_2}\left\{f^{(2)}(\xi)-f^{(2)}(x)\right\}(v-x)^2\left\{\frac{v^z\exp(-v)}{\gamma(z+1,a_2)-\gamma(z+1,a_1)}\right\}dv$$

is the remainder term with $\xi=\alpha(bv)+(1-\alpha)x$ for some $\alpha\in(0,1)$.

Observe that $a_1<z<a_2$ and $a_1,a_2,z\to\infty$ as $n\to\infty$. It follows from putting $(\beta_1,\beta_2,\beta_3)=(a_1,z,a_2)$ in Lemma 5.1 that

$$\frac{\gamma(z+2,a_2)-\gamma(z+2,a_1)}{\gamma(z+1,a_2)-\gamma(z+1,a_1)}-z$$
$$=1+o(1),\quad\text{and}$$
$$\frac{\gamma(z+3,a_2)-\gamma(z+3,a_1)}{\gamma(z+1,a_2)-\gamma(z+1,a_1)}-2z\frac{\gamma(z+2,a_2)-\gamma(z+2,a_1)}{\gamma(z+1,a_2)-\gamma(z+1,a_1)}+z^2$$
$$=z+2+o(1)$$
$$=\frac{x}{b}+2+o(1).$$

A similar argument to the Proof of Proposition 1 in Funke and Hirukawa (2017b) can also establish that $R_{\hat{f}_{(c_1,c_2)}(x)}=o(b)$. Then, the bias approximation immediately follows.

**Variance.** By

$$Var\left\{\hat{f}_{(c_1,c_2)}(x)\right\}=\frac{1}{n}E\left\{K_{G(x,b;c_1,c_2)}^2(X_i)\right\}+O(n^{-1}),$$

we only need to approximate $E\left\{K^2_{G(x,b;c_1,c_2)}(X_i)\right\}$. A straightforward calculation and the change of variable $w := 2u/b$ yield

$$
\begin{aligned}
&E\left\{K^2_{G(x,b;c_1,c_2)}(X_i)\right\} \\
&= \int_{c_1}^{c_2} \frac{u^{2x/b}\exp(-2u/b)\,f(u)}{b^{2(x/b+1)}\{\gamma(x/b+1,c_2/b)-\gamma(x/b+1,c_2/b)\}^2}\,du \\
&= \left\{\frac{b^{-1}\Gamma(2z+1)}{2^{2z+1}\Gamma^2(z+1)}\right\}\left\{\frac{\gamma(2z+1,2a_2)}{\Gamma(2z+1)}-\frac{\gamma(2z+1,2a_1)}{\Gamma(2z+1)}\right\} \\
&\quad \times \left\{\frac{\gamma(z+1,a_2)}{\Gamma(z+1)}-\frac{\gamma(z+1,a_1)}{\Gamma(z+1)}\right\}^{-2} \\
&\quad \times \int_{2a_1}^{2a_2} f\left(\frac{bw}{2}\right)\left\{\frac{w^{2z}\exp(-w)}{\gamma(2z+1,2a_2)-\gamma(2z+1,2a_1)}\right\}dw,
\end{aligned}
$$

where $(a_1, a_2, z) = (c_1/b, c_2/b, x/b)$ as in the bias approximation part, and the object inside brackets in the integral on the right-hand side is again a pdf on the interval $(2a_1, 2a_2)$. As before, the integral part can be approximated by $f(x) + O\left(b^{1/2}\right)$. By a similar argument to the one in the Proof of Lemma 5.2,

$$
\begin{aligned}
&\left\{\frac{b^{-1}\Gamma(2z+1)}{2^{2z+1}\Gamma^2(z+1)}\right\}\left\{\frac{\gamma(2z+1,2a_2)}{\Gamma(2z+1)}-\frac{\gamma(2z+1,2a_1)}{\Gamma(2z+1)}\right\} \\
&\quad \times \left\{\frac{\gamma(z+1,a_2)}{\Gamma(z+1)}-\frac{\gamma(z+1,a_1)}{\Gamma(z+1)}\right\}^{-2} \\
&= \frac{b^{-1/2}}{2\sqrt{\pi}x^{1/2}}+o\left(b^{-1/2}\right).
\end{aligned}
$$

Therefore, the variance approximation is also demonstrated, which completes the proof. ∎

# References

Ahmad, I.A., and Q. Li. 1997. Testing symmetry of unknown density functions by kernel method. *Journal of Nonparametric Statistics* 7: 279–293.

Aït-Sahalia, Y. 1996. Testing continuous-time models of the spot interest rate. *Review of Financial Studies* 9: 385–426.

Bagai, I., and B. L. S. Prakasa Rao. 1995. Kernel type density estimates for positive valued random variables. *Sankhyā: The Indian Journal of Statistics, Series A*, 57: 56-67.

Chen, S.X. 2000. Probability density function estimation using gamma kernels. *Annals of the Institute of Statistical Mathematics* 52: 471–480.

Engle, R.F., and J.R. Russell. 1998. Autoregressive conditional duration: a new model for irregularly spaced transaction data. *Econometrica* 66: 1127–1162.

Fan, Y., and A. Ullah. 1999. On goodness-of-fit tests for weakly dependent processes using kernel method. *Journal of Nonparametric Statistics* 11: 337–360.

Fé, E. 2014. Estimation and inference in regression discontinuity designs with asymmetric kernels. *Journal of Applied Statistics* 41: 2406–2417.

Fernandes, M., and J. Grammig. 2005. Nonparametric specification tests for conditional duration models. *Journal of Econometrics* 127: 35–68.

Fernandes, M., E.F. Mendes, and O. Scaillet. 2015. Testing for symmetry and conditional symmetry using asymmetric kernels. *Annals of the Institute of Statistical Mathematics* 67: 649–671.

Funke, B., and M. Hirukawa. 2017a. Nonparametric estimation and testing on discontinuity of positive supported densities: a kernel truncation approach. *Econometrics and Statistics*, forthcoming.

Funke, B., and M. Hirukawa. 2017b. Supplement to "Nonparametric estimation and testing on discontinuity of positive supported densities: a kernel truncation approach", *Econometrics and Statistics* Supplementary Data S1: https://doi.org/10.1016/j.ecosta.2017.07.006.

Guillamón, A., J. Navarro, and J.M. Ruiz. 1999. A note on kernel estimators for positive valued random variables. *Sankhyā: The Indian Journal of Statistics, Series A* 61: 276–281.

Hirukawa, M. 2010. Nonparametric multiplicative bias correction for kernel-type density estimation on the unit interval. *Computational Statistics and Data Analysis* 54: 473–495.

Hirukawa, M., and M. Sakudo. 2016. Testing symmetry of unknown densities via smoothing with the generalized gamma kernels. *Econometrics* 4: Article No.28.

Kulasekera, K.B., and J. Wang. 1998. Bandwidth selection for power optimality in a test of equality of regression curves. *Statistics and Probability Letters* 37: 287–293.

McCrary, J. 2008. Manipulation of the running variable in the regression discontinuity design: a density test. *Journal of Econometrics* 142: 698–714.

Otsu, T., K.-L. Xu, and Y. Matsushita. 2013. Estimation and inference of discontinuity in density. *Journal of Business and Economic Statistics* 31: 507–524.

Silverman, B.W. 1986. *Density Estimation for Statistics and Data Analysis*. London: Chapman and Hall.

# Chapter 6
# Asymmetric Kernels in Action

The final chapter presents two applications of asymmetric kernel smoothing to real data. One is on density estimation and the other on a testing problem. Each empirical illustration is closely related to the author's latest work.

## 6.1 Estimation of Income Distributions

As the first application, a variety of density estimates are compared. The data set of US annual family incomes in Abadie (2003), which has been already used to draw Fig. 1.1, is again employed for this analysis. The data are extracted originally from the 1991 Survey of Income and Program Participation, and 9275 family incomes are reported in thousands of US dollars. Table 6.1 provides a brief summary of the data.

**Table 6.1** Summary statistics on US family incomes

| # of Obs. | Mean | Std. Dev. | Skewness | Min. | Median | Max. |
|---|---|---|---|---|---|---|
| 9275 | 39.25 | 24.09 | 1.60 | 10.01 | 33.29 | 199.04 |

The following five density estimators are examined: (i) a parametric density estimator based on fitting $G(\alpha, \beta)$ by ML [ML]; (ii) the HG estimator in Sect. 3.3.3 with $G(\alpha, \beta)$ chosen as the parametric start and the G kernel used for bias correction [HG-G]; (iii) the G estimator [G]; (iv) the JLN-MBC estimator in Sect. 3.2.2 using the G kernel [JLN-G]; and (v) the JSH-MBC estimator in Sect. 3.3.3 with $G(\alpha, \beta)$ chosen as the parametric start and the G kernel used for bias correction [JSH-G]. The MLEs of $(\alpha, \beta)$ are $\left(\hat{\alpha}, \hat{\beta}\right) = (3.16, 12.43)$. To implement estimators (ii)-(v), we adopt the plug-in smoothing parameters proposed by Hirukawa and Sakudo (2014). Specifically, $\hat{b}_{GR-BU}$ are chosen for (ii) and (iii) and $\hat{b}_{GR-JLN}$ for (iv) and (v), where for

© The Author(s) 2018
M. Hirukawa, *Asymmetric Kernel Smoothing*, JSS Research Series
in Statistics, https://doi.org/10.1007/978-981-10-5466-2_6

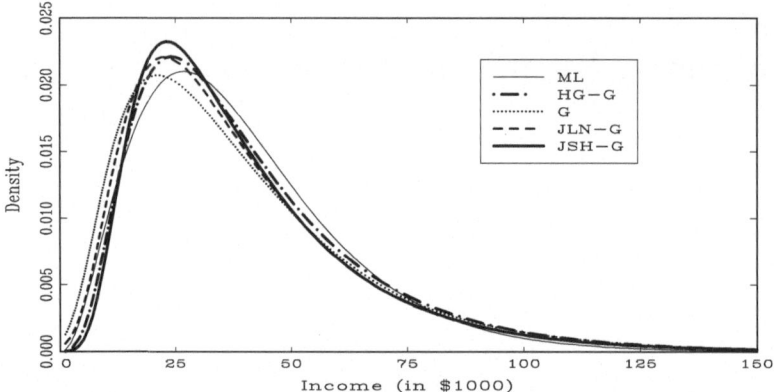

**Fig. 6.1** Estimates of the density of US family incomes

each formula MLEs $\left(\hat{\alpha}, \hat{\beta}\right)$ are plugged in place of $(\alpha, \beta)$. Notice that $\hat{b}_{GR-BU}$ is a special case of $\hat{b}_{GG}$ in Chap. 2.

Figure 6.1 presents plots of five density estimates. Although it is impossible to judge which estimator is closest to the truth, some interesting results can be observed. There is no substantial difference in the tail part. However, ML and G identify the position of the mode differently. Invoke that HG-G and JLN-G correct biases of ML and G, respectively. It appears that by downweighting (upweighting) the overestimated (underestimated) parts, HG-G and JLN-G shift the location of the mode to the left and right, respectively. As a result, their estimated modes become close in terms of both location and height. Since JSH-G is a bias-corrected version of HG-G, it yields a sharper peak through further downweighting the part in the vicinity of the origin.

## 6.2   Estimation and Testing of Discontinuity in Densities

The second application is concerned with estimation and testing procedures of discontinuity at multiple cutoffs. We employ the data sets on Israeli elementary schools used by Angrist and Lavy (1999), which contain enrollment counts of fourth and fifth graders at 2059 and 2029 schools, respectively. Following Maimonides' rule, Israeli public schools make each class size not greater than 40. As a result of strategic behavior on schools' and/or parents' sides, the density of school enrollment counts for each grade may be discontinuous at multiples of 40, namely 40, 80, 120, and 160. In fact, Otsu et al. (2013) and Funke and Hirukawa (2017) detect discontinuity (and thus evidence of manipulation in RDD) at some of these prespecified cutoffs.

Naturally, we are motivated to apply the joint discontinuity test at multiple cutoffs in Sect. 5.3 to Angrist and Lavy's (1999) data sets. On the other hand, it is desirable to clarify finite-sample properties of test statistics before analyzing the real data.

Therefore, this section conducts a small Monte Carlo study and then proceeds to the data analysis.

## 6.2.1  Finite-Sample Properties of Test Statistics for Discontinuity at Multiple Cutoffs

### 6.2.1.1  Monte Carlo Setup

The pdf of the Weibull distribution $W(1.75, 3.5)$ is used as the true density. To make the setup close to the real data, we examine the case with four suspected discontinuity points (i.e., $L = 4$). Specifically, 20, 40, 60, and 80% quantiles of the distribution are chosen as four cutoffs so that $\mathbf{c} = (c_1, c_2, c_3, c_4) :=$ $(q_{0.2}, q_{0.4}, q_{0.6}, q_{0.8})$. Let $X$ be drawn with probability $\gamma$ from the truncated Weibull distribution with support on $[0, q_{0.4})$ and with probability $1 - \gamma$ from the one with support on $[q_{0.4}, \infty)$. Unless $\gamma = \Pr(X < q_{0.4})$, the Weibull pdf is discontinuous at $q_{0.4}$. Also denote the measure of discontinuity as $d := \Pr(X < q_{0.4}) - \gamma$, where $d \in \{0.00, 0.02, 0.04, 0.06, 0.08, 0.10\}$ and $d > 0$ suggests a jump of the pdf at $q_{0.4}$. Two test statistics $T_i(\mathbf{c}) := nb^{1/2}\widetilde{\mathbf{J}}(\mathbf{c})^\top \widetilde{\mathbf{V}}_i(\mathbf{c})^{-1}\widetilde{\mathbf{J}}(\mathbf{c})$ for $i = 1, 2$ are investigated for $\widetilde{\mathbf{V}}_1(\mathbf{c})$ and $\widetilde{\mathbf{V}}_2(\mathbf{c})$ defined in Sect. 5.3, and three different values of the mixing exponent $\delta$ are considered, namely $\delta \in \{0.49, 0.64, 0.81\}$. For each test statistic, empirical rejection frequencies of the null $H_0 : \mathbf{J}(\mathbf{c}) = \mathbf{0}$ against the nominal 5 and 10% levels are computed. The sample size is 2000, which is also close to those of Angrist and Lavy's (1999) data sets, and 5000 replications are drawn.

The smoothing parameter $b$ for each test statistic is selected by the test-optimal criterion in Sect. 5.4. Let $n_\ell$ be the number of observations falling into the interval $[c_\ell, c_{\ell+1})$ for $\ell = 0, \ldots, 4$, where $(c_0, c_5) \equiv (0, \infty)$. Then, implementation details are summarized as the following five steps.

**Step 1:** Pick $M := \left\lfloor \min_{0 \le \ell \le 4} n_\ell^p \right\rfloor$ for $p = 1/2$.

**Step 2:** Make $M$ subsamples of sample size $k := \sum_{\ell=0}^4 k_\ell$, where $k_\ell := \lfloor n_\ell/M \rfloor$ for $\ell = 0, \ldots, 4$.

**Step 3:** Choose the interval for $b_k$ as $H_k := [0.15, 0.50]$.

**Step 4:** Set the critical value at $\chi^2_{0.05}(4) = 9.49$ and find
$$\hat{b}_k = \inf \left\{ \arg\max_{b_k \in H_k} \hat{\pi}_M(b_k) \right\} \text{ by a grid search,}$$
where $\hat{\pi}_M(b_k) := (1/M) \sum_{m=1}^M \mathbf{1}\left\{ T_m(\mathbf{c}) > \chi^2_{0.05}(4) \right\}$
for the test statistic from the $m$th subsample $T_m(\mathbf{c})$.

**Step 5:** Recover $\hat{B} = \hat{b}_k k^q$ for $q = 4/9$ and obtain $\hat{b}_n = \hat{B} n^{-q}$.

**Table 6.2** Empirical rejection frequencies of test statistics

| Test | $\delta$ | Nominal | $d$ (%) | | | | | |
|---|---|---|---|---|---|---|---|---|
| | | | 0.00 | 0.02 | 0.04 | 0.06 | 0.08 | 0.10 |
| $T_1$ (c) | 0.49 | 5% | 4.2 | 5.8 | 10.8 | 23.5 | 45.3 | 70.6 |
| | | 10% | 8.6 | 10.8 | 19.0 | 35.8 | 60.1 | 81.0 |
| | 0.64 | 5% | 4.2 | 5.7 | 10.4 | 22.6 | 43.5 | 67.1 |
| | | 10% | 9.0 | 11.0 | 18.8 | 34.6 | 57.6 | 78.4 |
| | 0.81 | 5% | 4.1 | 5.8 | 10.3 | 21.9 | 41.6 | 64.4 |
| | | 10% | 9.3 | 11.0 | 18.6 | 34.2 | 55.4 | 76.2 |
| $T_2$ (c) | 0.49 | 5% | 4.8 | 6.3 | 11.3 | 24.8 | 46.6 | 71.4 |
| | | 10% | 9.7 | 11.6 | 20.3 | 36.9 | 61.4 | 81.7 |
| | 0.64 | 5% | 4.6 | 6.3 | 11.0 | 23.8 | 44.5 | 68.2 |
| | | 10% | 9.7 | 11.9 | 19.7 | 36.2 | 58.8 | 79.2 |
| | 0.81 | 5% | 4.6 | 6.5 | 11.0 | 23.0 | 43.1 | 65.4 |
| | | 10% | 10.2 | 12.0 | 19.9 | 35.5 | 56.6 | 76.9 |

#### 6.2.1.2 Results

Table 6.2 reports simulation results. Observe that each test statistic has good size ($d = 0$) and power ($d > 0$) properties. As with other tests smoothed by asymmetric kernels, the good performance is based solely on first-order asymptotic results. A closer look also reveals that $T_2$ (c) has better power than $T_1$ (c) for each $\delta$ and that $\delta = 0.49$ is most powerful.

### 6.2.2 Empirical Illustration

We proceed to applying the joint discontinuity test to Angrist and Lavy's (1999) data sets. $T_2$ (c) with $\delta = 0.49$ with the test-optimal smoothing parameter plugged in is employed because of its better finite-sample properties. The test statistic takes the values of 38.58 and 37.69 for fourth and fifth graders, respectively. Each value well exceeds the critical value at the 1% level $\chi^2_{0.01}(4) = 13.28$, i.e., there is strong evidence that the underlying density for each grade is discontinuous at some of four prespecified cutoffs.

Moreover, Funke and Hirukawa (2017) conduct discontinuity tests at the four cutoffs individually for each grade and find evidence of discontinuity at 40, 120, and 160 for fourth graders and at 40 and 120 for fifth graders. Figure 6.2 plots estimates of enrollment densities for fourth (solid) and fifth graders (dashed) under the assumption that the densities are indeed discontinuous at all these cutoffs. A difficulty arises in smoothing parameter selection. While there is no guidance for implementing density estimation with discontinuity points, adopting the smoothing parameter value under

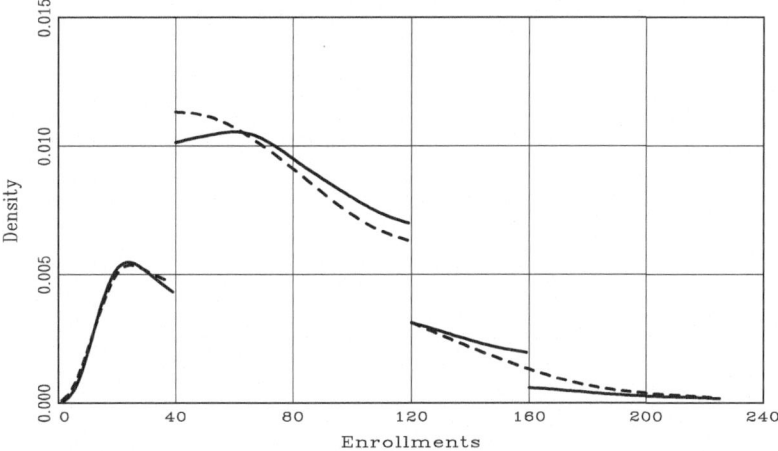

**Fig. 6.2** Density estimates of school enrollments in the presence of multiple cutoffs

the test-optimal criterion is inappropriate in that this is an estimation problem. As a consequence, we simply put $\hat{b} = \hat{\sigma} n^{-2/5}$ for each density estimate, where $n$ is the sample size (i.e., $n = 2059$ for fourth and $n = 2029$ for fifth graders) and $\hat{\sigma}$ the sample standard deviation.

# References

Abadie, A. 2003. Semiparametric instrumental variable estimation of treatment response models. *Journal of Econometrics* 113: 231–263.

Angrist, J.D., and V. Lavy. 1999. Using Maimonides' rule to estimate the effect of class size on scholastic achievement. *Quarterly Journal of Economics* 114: 533–575.

Funke, B., and M. Hirukawa. 2017. Nonparametric estimation and testing on discontinuity of positive supported densities: a kernel truncation approach. *Econometrics and Statistics*, forthcoming.

Hirukawa, M., and M. Sakudo. 2014. Nonnegative bias reduction methods for density estimation using asymmetric kernels. *Computational Statistics and Data Analysis* 75: 112–123.

Otsu, T., K.-L. Xu, and Y. Matsushita. 2013. Estimation and inference of discontinuity in density. *Journal of Business and Economic Statistics* 31: 507–524.

# Index

**A**
Absolute regularity, *see* $\beta$-mixing
ACD model, 73
Actuarial loss distribution, 7
Additive bias correction, 42, 43
$\alpha$-mixing, 30, 64

**B**
Beta function, 8
Beta kernel, 5, 8, 11, 20, 29, 54, 61
$\beta$-mixing, 31
Bias-variance trade-off, 23, 86
Birnbaum–Saunders kernel, 5, 30
Boundary bias, 1, 2, 4
Boundary correction, 2

**C**
Cross-validation
  $h$-block cross-validation, 36, 70
  leave-one-out cross-validation, 36

**D**
Degenerate $U$-statistic, 80
Diffusion estimator, 65
Diffusion model, 7, 64
Digamma function, 43, 44
Distribution of trading volumes, 7
Drift estimator, 65
D-test, 75

**G**
Gamma function, 7, 8

**Gamma kernel**, 5, 8, 11, 29, 43, 52, 61, 65,
  75, 78, 85, 103
Gaussian kernel, 24, 25, 62
Gaussian-copula kernel, 5, 31
Generalized Birnbaum-Saunders kernel, 5
Generalized cross-validation, 70
Generalized gamma kernel, 5, 9, 11, 20, 22,
  24, 29, 34
Generalized inverse Gaussian kernel, 5
Generalized jackknife method, 43
General weight function estimator, 3

**H**
Hazard rate, 7, 30, 75
H-test, 75

**I**
Income distribution, 6, 7, 103
Integrated squared error, 35, 79
Inverse gamma kernel, 5
Inverse Gaussian kernel, 5, 30

**J**
Joint density estimation, 32, 50

**K**
Kernel regression estimator
  Gasser-Müller estimator, 3, 63
  local linear estimator, 59
  Nadaraya–Watson estimator, 59, 65

**L**
Legendre duplication formula, 37

© The Author(s) 2018
M. Hirukawa, *Asymmetric Kernel Smoothing*, JSS Research Series
in Statistics, https://doi.org/10.1007/978-981-10-5466-2